Analytical Affinity Chromatography

Editor

Irwin M. Chaiken

National Institute of Diabetes and Digestive
and Kidney Diseases
National Institutes of Health
Bethesda, Maryland

CRC Press
Taylor & Francis Group
Boca Raton London New York

CRC Press is an imprint of the
Taylor & Francis Group, an **informa** business

First published 1987 by CRC Press
Taylor & Francis Group
6000 Broken Sound Parkway NW, Suite 300
Boca Raton, FL 33487-2742

Reissued 2018 by CRC Press

© 1987 by CRC Press, Inc.
CRC Press is an imprint of Taylor & Francis Group, an Informa business

Library of Congress Cataloging-in-Publication Data

Analytical affinity chromatography.

   Includes bibliographies and index.
   1. Affinity chromatography.   2. Biomolecules—
Analysis.   I. Chaiken, Irwin M.   [DNLM:
1. Chromatography, Affinity.   QU 25 A532]
QP519.9.A35A53   1987   574.19'285   86-28410
ISBN 0-8493-5658-X

A Library of Congress record exists under LC control number: 86028410

Publisher's Note
The publisher has gone to great lengths to ensure the quality of this reprint but points out that some imperfections in the original copies may be apparent.

Disclaimer
The publisher has made every effort to trace copyright holders and welcomes correspondence from those they have been unable to contact.

ISBN 13: 978-1-315-89057-9 (hbk)
ISBN 13: 978-1-351-06967-0 (ebk)

Visit the Taylor & Francis Web site at http://www.taylorandfrancis.com and the
CRC Press Web site at http://www.crcpress.com

PREFACE

Specific surface recognition of folded macromolecules is a recurrent theme in biology. This fact has stimulated the development of affinity chromatography as a powerful separation method to purify biological marcomolecules and small molecules by attaching a known molecular interactor to an insoluble matrix and then adsorbing a soughtafter molecular species, multimolecular assembly, or even cell type from mixtures eluted through the affinity matrix. The use of immobilized starch by Emil Starkenstein in Prague (1910) to separate amylase was, no doubt, an incisive progenitor of modern affinity chromatographic purification. And, the studies of cellulose-bound enzymes by Ephraim Katchalski in Israel, in the 1950's and 60's, emphasized the functional viability of immobilized biomolecules. Yet it was in the late 1960's, with the development of agarose matrix chemistry in Sweden by Rolf Axén, Lars Sundberg, and Jerker Porath; the use of agarose-bound ligands to purify enzymes in the United States by Pedro Cuatrecasas, Meir Wilchek, and Christian Anfinsen; and an elevated understanding of macromolecular structure in general, that the explosive growth of the affinity chromatographic method started in earnest.

By the early 1970's, the growing force of its reliability as a biospecific separation method led to an awareness that affinity chromatography also had potential use as an analytical method to study mechanisms of the interactions between mobile and immobilized biomolecules. Thus, if a mobile molecule binds to an affinity matrix through the complementary binding surface of the immobilized molecule but not with other nonbiospecific matrix sites, affinity chromatographic elution behavior of the mobile form can be taken as a direct measure of quantitative interaction properties of mobile and immobilized interactants, including equilibrium binding (or dissociation) constants and, within more restricted circumstances, rate constants for association and dissociation. Using competitive elution, with mobile effectors that compete with immobilized species for binding to mobile interactants, binding properties for the solution interaction also can be determined. Affinity chromatographic analysis of molecular interactions is not limited by the size of the mobile or immobilized interactors (vs the size constraints of ultracentrifugation, equilibrium dialysis, or gel filtration), allows a direct experimental measure of an interaction process (vs the need, as in spectroscopy, to infer interactions from the effects of interaction), is not dependent on a particular functional activity (as in enzyme kinetics and immunoassay), and can be performed on a micro (down to picograms or less if detection methods permit) as well as macro scale of both affinity column and mobile interactors (vs the more restricted concentration demands of many of the above methods). The benefits of characterizing macromolecular interactions using immobilized molecules have stimulated a steady development of analytical uses for affinity chromatography. The methodology has been referred to as "quantitative affinity chromatography" to emphasize the quantitative approach and the nature of information obtained. Yet, to acknowledge the more general usefulness of the method to study binding both qualitatively and quantitatively, we have chosen to adopt the more comprehensive name "analytical affinity chromatography".

This volume presents discussions of theoretical and experimental considerations that have led to the analytical affinity chromatography field, as well as current efforts to use this methodology to characterize the interaction mechanisms of biological macromolecules and to establish conditions for employing bioaffinity chromatographic systems as preparative tools. The chapters include a comprehensive discussion of interactive chromatography theory (DeLisi and Hethcote), a review of experimental data obtained for biological macromolecules and the relevant theoretical considerations of affinity chromatography which led to them (Swaisgood and Chaiken), an evaluation of rate processes in affinity chromatography and the potential to determine biologically meaningful chemical rate constants (Walters), and the use of quantitative and molecular considerations to design affinity chromatographic systems (Stellwagen and Liu). These topics have been chosen to reflect major trends of thinking and experimental work, both historical and current. Quite clearly, all of the chapters reflect an optimism that affinity chromatography offers an effective and yet experimentally straightforward way to measure biomolecular interaction properties. This optimism is borne of the overall finding that binding characteristics determined chromatographically, especially equilibrium constants, are almost always equivalent or closely similar to those determined fully in solution (when the latter are available for comparison). Nonetheless, attention has been given in the texts to cite not only the strengths but also the limitations and uncertainties of the methodology. The reasons for observed differences between chromatographic and solution binding parameters, including nonspecific components of the affinity system and binding characteristics specific for a particular interactant in an immobilized form, are addressed where possible.

Development of the analytical affinity chromatography field has benefited from the work of scientists from several disciplines, including biochemistry, physical and analytical chemistry, and mathematics. Such multidisciplinary contributions have been extremely helpful in critically examining the theoretical and experimental validity of analytical affinity chromatography as a biochemical method. As a set, the chapters provide a composite view of the varied approaches which have been brought to the method; however, the chapters do vary significantly in content and style and therefore have been organized so that each can be read in isolation, depending on the interests of the reader, with appropriate cross referencing where needed. Multidisciplinary input has had one major drawback, the usage of several sets of symbols to represent chromatographic and interaction characteristics. In this book, the reader will find a glossary of terms at the end of each chapter. In addition, an appendix is included after the four chapters to correlate the various notations used. At least at the current stage of development and use, it is likely that multiple notations will persist in the literature in general; it is hoped that the comparisons provided here will help the reader to understand the interrelationships of theory and to unify his or her appreciation of the state of the field.

The quantitative use of affinity chromatography has been developed sufficiently that it is now used increasingly both to characterize molecular interactions when the interaction mechanisms themselves are of primary interest and to characterize binding constraints and performance of affinity matrices when their use for preparative purification is the major goal. Given the potential to apply high-performance liquid chromatography and accompanying miniaturization to affinity systems, the prospect of using the analytical approach to develop affinity columns for microdiagnostics in both research laboratories and in clinics also seems realistic. As the overall technology of affinity-based separation using immobilized biomolecules expands and develops, analytical affinity chromatography promises to continue evolving as a flexible general tool for both biomedical research and biotechnology.

EDITOR

Irwin Chaiken, Ph.D., is a senior investigator at the National Institutes of Health, in the National Institute of Diabetes and Digestive and Kidney Diseases. Dr. Chaiken was born in Rhode Island, attended Classical High School in Providence, and received his undergraduate education at Brown University, where he earned a Bachelor of Arts degree in Chemistry in 1964. He received a Ph.D. degree in Biological Chemistry in 1968 at the University of California, Los Angeles, under the tutelage of Prof. Emil L. Smith. After research studies as a postdoctoral scholar at UCLA and as a postdoctoral fellow at the National Institutes of Health working with Dr. Christian B. Anfinsen, he joined the NIH staff in 1970, first as a staff fellow and, from 1972, as a senior investigator.

Concurrently, Dr. Chaiken has had a long-standing cooperative reserach association and exchange with the Institute of Organic Chemistry of the University of Padua in Italy; was an adjunct faculty member in the Chemistry Department at the University of Maryland from 1973 to 1981; was an exchange scientist of the U.S. National Academy of Science in Prague during September 1985; and was a visiting professor in the Biophysics Department at Johns Hopkins University Medical School, 1985 to 1986. Starting in 1987, he is an adjunct faculty member of the Biochemistry Department, Georgetown University Medical School. He has taught courses at the University of Maryland and at the National Institutes of Health on the structure, interactions, function, and design of peptides and proteins and currently teaches both basic and biotechnology courses on this subject.

Dr. Chaiken has served as chairman of the 5th International Symposium on Affinity Chromatgraphy and Biological Recognition in 1983 in Annapolis. Since 1983 he has served as an organizer and, since 1985, as president of the International Interest Group in Biorecognition Technology. He serves as Editor-in-Chief of the IIGBT Science Newsletter. Since 1985 he has served on the long-range planning committee of the American Peptide Symposium and is Chairman-elect of the Gorden Research Conference on Immobilized Systems in Biotechnology, 1988/1990.

Dr. Chaiken's main areas of research interest are macromolecular recognition and biorecognition technology; peptide and protein engineering; and multimolecular assembly of peptides and proteins in biological systems. He has authored many primary research publications and reviews, has lectured internationally and has refereed for many scientific journals. He is a member of the American Society of Biological Chemists, the American Association for the Advancement of Science, the New York Academy of Science, Sigma Xi, and the International Interest Group in Biorecognition Technology.

CONTRIBUTORS

Irwin M. Chaiken
Senior Investigator
National Institute of Diabetes and
  Digestive and Kidney Diseases
National Institutes of Health
Bethesda, Maryland

Charles DeLisi
Associate Director for Health and
  Environmental Research
Office of Energy Research
Department of Energy
Washington, D.C.

Herbert W. Hethcote
Professor
Department of Mathematics
University of Iowa
Iowa City, Iowa

Yin-Chang Liu
Research Assistant
Department of Biochemistry
University of Iowa
Iowa City, Iowa

Earle Stellwagen
Professor
Department of Biochemistry
University of Iowa
Iowa City, Iowa

Harold E. Swaisgood
William Neal Reynolds Professor of Food
  Science and Biochemistry
Department of Food Science
North Carolina State University
Raleigh, North Carolina

Rodney R. Walters
Research Scientist
The Upjohn Company
Drug Metabolism Research
Kalamazoo, Michigan

TABLE OF CONTENTS

Chapter 1

CHROMATOGRAPHIC THEORY AND APPLICATION TO QUANTITATIVE AFFINITY CHROMATOGRAPHY

Charles DeLisi and Herbert W. Hethcote

TABLE OF CONTENTS

I. INTRODUCTION

Liquid column chromatography and affinity chromatography are widely used for separating and purifying macromolecules.[1-4] Chromatography is much less widely used as a quantitative method for studying chemical reactions, i.e., for estimating equilibrium and rate constants.[5] In this chapter we establish a firm theoretical foundation for quantitative affinity chromatography by giving precise mathematical descriptions of the movements of molecules in a chromatography column and of the interactions of molecules with free ligands and with ligands attached to the beads. The mathematical descriptions lead to formulas for the moments of elution profiles (e.g., the mean and the variance). These formulas can be used with experimental data to estimate chemical reaction equilibrium and rate constants.

Many aspects of chromatography are considered. Formulas for means and variances of elution profiles are obtained for both small- and large-zone chromatography. If the flow rate is large enough to satisfy some given inequalities, then the effects of diffusion on the moments of the elution profiles can be neglected. Expressions for the height equivalent to a theoretical plate are obtained from formulas for the mean and variance, and the peak of the elution profile is considered as an approximation to the mean. The effects of heterogeneities in the molecules and in the beads are analyzed.

Affinity chromatography is considered with either the ligands or the macromolecules attached covalently to either porous or impenetrable beads. The macromolecule-ligand binding can be either monovalent or bivalent, and the chemical reactions can be either fast or slow compared to the mass transfer kinetics in and out of the beads. Special sections outline methods for estimating equilibrium and rate constants. The last section summarizes all of the chromatography and affinity chromatography formulas. This chapter not only contains a unified presentation of our theory of chromatography and affinity chromatography as developed in a sequence of papers,[6-10] but it also contains several new results.

Quantitative studies of the reaction between solution-phase molecules and molecules bound to beads in a column are assuming increasing importance because the reactions are analogous to those between ligand- and cell-bound receptors. The latter are known to play an important role in the regulation of cellular activity;[11] however, simply knowing that a ligand-receptor reaction must occur is not sufficient to understand the physical basis of regulation. Indeed, cells may be triggered to different types of activity merely by changing physical parameters such as receptor valence, affinity, and number.[12,13] Moreover, binding often involves multiple steps that may trigger competitive processes, and when this happens, knowledge of the rate constants for each step is important.

In addition, receptors are sometimes heterogeneous in their affinity for ligands, as in the case of B-cell-bound immunoglobulin, and the distribution must be known if one wishes to understand how antibody affinity is regulated. Finally, theory predicts that the equilibrium and rate constants for cell-bound receptors may differ substantially from their values when the receptors are dispersed in solution[14] and in the case of rate constants, this difference can be present even when the reaction mechanism is the same. Systematic experimental studies of the relation between the two will be of considerable importance if we wish to reliably extrapolate — as must often be done — from dispersed to cell-bound systems.

In conventional methods, such as dialysis, the reactive partners are both uniformly dispersed in solution. In affinity chromatography there is not only a reaction between the eluted macromolecules and solution-phase ligands, but also a reaction between the macromolecules and ligands covalently attached to the beads. Thus with chromatography methods these two distinct chemical reactions can be studied simultaneously and their equilibrium and rate constants can be compared. Although several different methods of varying degrees of complexity are available for estimating constants and their distributions, it is evident that a widely available, simple, fast, and reliable method for obtaining all of this information simultaneously would be very useful. Affinity chromatography has great potential as such a quantitative method.

II. LIQUID COLUMN CHROMATOGRAPHY

A. The Molecular Basis of Recognition
1. Introduction

This section is concerned primarily with the theoretical basis for determining thermodynamic and kinetic constants governing macromolecular interactions. The characterization of interactions by these parameters is phenomenological, i.e., although the values of rate and equilibrium constants are determined by the details of molecular interactions, they do not explicitly reflect these details. In this section we discuss in an approximate, intuitive, and semiquantitative manner, the connection between rate constants and the molecular environment of the interaction. We begin by reviewing the phenomenology of rate constants and the relative contributions of motion and chemical reaction to their values first in a uniform mixture in which motion is diffusive and then in a convective environment. We then discuss the relation between the reactive part of the rate constant and the forces that exist between beads and molecules in a chromatography experiment.

2. Reactive, Diffusive, and Transport Effects on Rate Constants
a. Diffusion Mediated Collision

To introduce the pertinent ideas we will begin by considering a very simple picture with two types of reactants, A and B, undergoing Brownian motion and occasionally colliding to form a product AB. Product formation requires not only a suitable distance between the centers of mass of the reactive units, but also their correct mutual orientation.[15] The ideas to follow can be presented most readily by neglecting explicit inclusion of orientation; a development that finds approximate realization in spherically symmetric molecules and in molecules in which orientational diffusion occurs very rapidly compared to center of mass diffusion. This picture leads to the intuitive expectation that the overall reaction-rate constant can be decomposed into two parts, one determining the rate at which the molecules collide to form an encounter complex, and another determining the rate at which the encounter complex reacts to form the product. We represent the overall process as

$$A + B \underset{k_-}{\overset{k_+}{\rightleftarrows}} A...B \underset{k_{-1}}{\overset{k_1^*}{\rightleftarrows}} AB \tag{1}$$

where k_+ and k_- are the (translational) diffusive-rate constants, and k_1^* and k_{-1} are reactive-rate constants, and it is understood that diffusional orientation is sufficiently rapid that orientation can be included as a multiplicative factor in k_1. One can show that with $k_{-1}/$

$k_i^* \ll 1$, this two-step process can be represented by a single-step process with effective forward- and reverse-rate constants:[16-18]

$$k_f = k_i^* k_+/(k_i^* + k_-) \tag{2}$$

and

$$k_r = k_- k_{-i}/(k_i^* + k_-) \tag{3}$$

The time constant governing the reaction is given in terms of k_f and k_r by $\tau = 1/(k_f C + k_r)$ where C is related to the concentration of reactants. The reactive-rate constants are determined by the molecular forces between the interacting units as discussed below; the diffusive-rate constants are determined by the macroscopic properties of the medium in which the reactants are moving.

A large literature exists on the calculation of diffusive-rate constants, dating from the pioneering work of Smoluchowski (see, for example, the review by Noyes[19]) to the new developments of Keiser.[20] Here we present a brief sketch intended only to introduce some key ideas.

For an infinite three-dimensional medium in which a single type of reaction is present at a very dilute concentration C, the mean time τ between collisions is

$$\tau = 1/(4\pi \, aDC) \tag{4}$$

where D is the sum of the diffusion coefficients of the two reactive units and "a" is the distance between their centers (often taken as the sum of their van der Waals radii) when they are in an encounter complex. The diffusive forward rate constant is

$$k_+ = 1/\tau C = 4\pi aD \tag{5}$$

To obtain the diffusive reverse rate constant k_- we note that the mean time for units in an encounter complex to double their separation is

$$\tau_r = a^2/3D \tag{6}$$

and hence an estimate for k_- is

$$k_- = 3D/a^2 \tag{7}$$

An essential aspect of chromatography is the interaction of macromolecules with beads or with ligands covalently attached to beads. If the interaction is with a large, uniformly reactive bead, the expressions for the rate constants are those given above. If, on the other hand, the reaction is with ligands uniformly distributed over an otherwise unreactive bead and occupying a small fraction of the bead surface, the expressions for the diffusive rate constants change in an important way. If the bead-bound ligand is treated as a disk of radius "s", and N ligands are uniformly distributed over a bead of radius "a", then[21]

$$k_+ = 4\pi aD \, Ns/(Ns + \pi a) \tag{8}$$

and[22]

$$k_- = \frac{3D}{\upsilon s^2} \frac{\pi a}{Ns + \pi a} \tag{9}$$

where υ is a factor of order unity.

If, in addition, the diffusing particle is under the influence of a central potential V(r), then[23,24]

$$k_+ = \frac{4\pi aD \; Ns \; \lambda^*}{\pi a + Ns\lambda^* exp(-V^*/kT)} \tag{10}$$

where

$$\lambda^* \equiv exp(-V(a)/kT)$$

and

$$exp(-V^*/kT) \equiv a \int_a^\infty (\lambda^{-1}(p)/p^2) \; dp$$

Under these conditions, the reverse rate constant is given by

$$k_- = \frac{3D}{vs^2} \frac{\pi a}{\pi a + Ns\lambda^* exp(-V^*/kT)} \tag{11}$$

In these equations, kT is the thermal energy. An interesting aspect of Equations 8 and 10 is that N need not be large for the diffusive forward-rate constant to be close to the value it would have if the sphere were uniformly reactive. Thus from Equation 8 with s = 30Å and a bead of radius 5×10^{-4} cm, the forward rate constant is at one half its maximum value with about 3000 ligands/bead.

b. Transport-Mediated Motion

In column chromatography, solute is transported by the flow of solvent under the influence of a gravitational field. If no beads were present, a small volume element not too close to the column boundary would follow a smooth flow line. If a round spherical object, such as a bead, is in the path of this line, the line will curve smoothly, following the contours of the boundary. Such laminar flow conditions prevail if the Reynold's number

$$R \equiv \rho va/\mu \tag{12}$$

is of order unity or smaller. In Equation 12, ρ is the fluid density, "v" is its velocity, μ is the coefficient of viscosity and "a" is some characteristic dimension of the boundary in the direction of flow, which we take to be the bead radius. Using a = 5×10^{-4} cm, v = 0.2 cm/sec and, for water, μ = 0.01 P, and ρ = 1 g/cm, the Reynold's number is 0.01.

Since the Reynold's number is the ratio of the inertial force of the moving fluid to the frictional drag force exerted by the bead (i.e., viscosity), it is evident that the latter dominates close to the bead surface. Thus, even if to a good approximation drag forces between the stream lines sufficiently far from the bead can be neglected, drag forces of the fluid on the boundaries cannot be completely ignored. The nature of these forces and their manipulation is central to developing an effective column, as we indicate below. For the present, we note that they act tangentially to the flow and tend to slow the velocity in the vicinity of the boundary. The velocity thus increases from zero at the surface, to a constant rate beyond some transition distance. Evidently, at the surface, only Brownian motion occurs, and to within some distance from the surface it will dominate transport. As we indicate in subsequent sections, however, for distances at which the full flow velocity is reached, transport is generally the dominant mode of movement in the direction parallel to the long axis of the column.

The above view of motion along the column suggests three different steps in the overall reactive process. Molecules move along flow lines, with Brownian movement between flow lines and across the transition layer surrounding a bead. Inside the transition layer, the molecule moves by diffusion. We also assume that within the bead, because of its tortuosity, molecules move more or less randomly. We therefore assume Brownian motion at all points within the transition layer. The diffusing molecule can form an encounter complex with a bead-bound ligand and the encounter complex can then form a chemical bond resulting in the product AB.

To our knowledge, a theory of chemical reactions in this environment, comparable to the theory presented in Section A.2.a has not been developed, nor is it our intention to develop such a theory here. Our purpose is only to present a physical picture, a framework for thinking about the overall reactive process, and to identify and estimate some of the more important parameters entering this framework.

We begin by estimating the distance from the surface at which the tangential drag forces between flow lines (i.e., viscous forces) become comparable to inertial forces of flow (for a detailed treatment see H. Schlichting[25]). This boundary layer thickness, which is typically estimated by assuming a linear spatial-velocity gradient orthogonal to the surface, is to be distinguished from the transition layer within which diffusion dominates transport.

The boundary layer thickness δ turns out to be

$$\delta \cong 5a/\sqrt{R} \tag{13}$$

where R is the Reynold's number associated with the bead of radius "a". Using the above values, we find that $\delta = 0.025$ cm, or about 25 times the bead diameter. Unless the column is very loosely packed, viscous forces will dominate at all positions and the flow lines will not be independent of one another. This result has implications for the general theory of chromatography, namely the transport velocity cannot be treated as a constant. How a constant velocity assumption, which is used throughout the remainder of this chapter, affects results is currently not known. It is likely, however, that an effective velocity which is some complicated, though explicitly determined, function of the full velocity profile is sufficient to mimic the elution profile.

The large value of δ indicates that it will not be a good candidate for demarcating the transition from Brownian motion to transport. An appropriate expression can, however, be obtained by relatively simple considerations. We show in Sections C.2.b and D.4 that, insofar as effects on the elution profile are concerned, the critical dimensionless parameter determining the relative importance of Brownian motion and flow is the ratio

$$r = vh/D \tag{14}$$

where "h" is the height of the column and D is the diffusion coefficient. At a local level, "h" must be replaced by the characteristic dimension of interest, in the present case, "a". Furthermore, if we continue to assume that the velocity gradient orthogonal to the surface is linear (which is now a better approximation than it was in the derivation of Equation 13), take $r = 1$ at the boundary of the transition region, and denote by the superscript * the values of variables at this boundary, then

$$v\delta^*/\delta = v^* = D/a \tag{15a}$$

or

$$\delta^* = D\delta/(va) \tag{15b}$$

The first equality in Equation 15a results from the assumption of a linear velocity profile between δ and $\delta*$. The second follows from the assumption that "r" is of order 1 at the transition from Brownian to transport dominated motion. With $D = 10^{-6}$ cm²/sec, we find that $\delta* = 2.5 \times 10^{-4}$ cm, or about a fourth of the bead diameter.

The physical picture of motion that emerges is the following. At the bead surface the forces of attraction bring the flow velocity to zero, and sufficiently near but off the surface, Brownian motion must dominate gravitational transport. Similarly, within the bead we assume that Brownian motion dominates, though for reasons related to the complex configurations of the polymers composing the bead. As distance out from the bead surface increases, so does the transport velocity, and at a distance $\delta*$, a transition occurs from the dominance of Brownian motion to the dominance of flow. Even after the distance $\delta*$ is exceeded, however, velocity flow lines continue to interfere with one another and never become fully independent because of the close proximity of neighboring beads as indicated by Equation 13.

We can use this picture to help us gain insight into the relations between the various parameters and the relative importance of their contributions to the overall reaction rate by formulating the ideas mathematically. It is not our intention to develop a detailed treatment here, but only to present heuristic arguments in order to identify key parameters. To this end we take all movement within $\delta*$ as diffusive, and all movement outside $\delta*$ as convective; the former characterized by a single diffusion coefficient D and the latter by a single-position independent transport velocity, "v". Evidently three different rates combine to determine the rate of the overall reaction of a solvent molecule with a bead or with a bead-bound molecule: the rate of arrival at $\delta*$ by transport, the rate of moving from $\delta*$ to the bead surface by diffusion, and the rate of reacting, once at the surface. Thus the overall reaction time T is

$$T = T_v + T_D + T_r \tag{16}$$

The first term on the right is the mean time for a ligand flowing down the column to arrive at a transition layer; the second is the mean time, once inside the transition layer, to form an encounter complex with a bead-bound molecule. The estimate of this time should take account of the possibility that a ligand in the transition layer may leave it before encountering a bead-bound ligand. The third term is the rate of reaction once an encounter complex is formed. Estimates of the first two times involve solving the equations governing the large-scale motion of ligands that will be central to this chapter; an estimate of the third[26,27] is a problem in quantum chemistry and is not treated here. If "d" is the mean distance between the boundaries of diffusive shells surrounding adjacent beads, then $T_v = d/4v$.

The mean time for a particle, having arrived within $\delta*$, to reach the surface by diffusion can be obtained by methods which will be central to the remainder of this section. We postpone their development and simply note that the mean time W associated with a particular rate constant, for example the forward rate, satisfies the following differential equation.

$$\frac{d^2W}{dr^2} - \frac{2}{r}\frac{dW}{dr} - u\frac{dW}{dr} = -H(r_1 - r)/D \tag{17}$$

where

$$u \equiv (D/kT) \, dV/dr$$

V is some central potential (e.g., of a van der Waals type) associated with the bead; r_1 is the distance, measured from the center of the bead, at which the particle begins its diffusive journey; and $H(x)$ is 1 for $x \geqslant 0$ and 0 otherwise. The solution to Equation 17, i.e., the time to reach "a", starting at $r_1 > a$ given a reflective boundary at $b > r$, is

$$W(a) = (1/D)\left\{ \int_{r_1}^{a} \exp(f(r)) \, dr \int_{r_1}^{a} \exp(-f(r')) \, dr' - \int_{r_1}^{a} \exp(-f(r)) \, dr \int_{r_1}^{b} \exp(f(r)) \, dr \right\} \quad (18)$$

where

$$f(r) \equiv \int (2/r + u/D) \, dr$$

The integrals can be easily evaluated for $u = 0$ or any other constant. The relevance of the latter can be seen by noting that the distance from the surface over which diffusive motion occurs is small compared to the distance from the bead center (the latter being the origin of V), so that the gradient of V should be relatively small. As an indication of what an explicit result involves, we write the solution for $u = 0$.

$$W(a) = a^2/6D - r_1^2/6D + b^3(r_1 - a)/3Dr_1a \quad (19)$$

Equation 19 is the appropriate expression for obtaining the diffusive forward-rate constant. If we want to make the connection with the Smoluchowski results for bulk solution, however, we need to proceed further and average Equation 19 over all possible starting positions. The result is[21]

$$\langle t \rangle = \frac{a^2}{6D} + \frac{b^3}{3aD} - \frac{1}{(b^3 - a^3) \, D} \left[\frac{(b^5 - a^5)}{10} + \frac{b^3(b^2 - a^2)}{2} \right] \quad (20)$$

If we take the reciprocal of Equation 20 and multiply by the volume $4/3 \, \pi b^3$ we obtain the diffusive forward rate constant.

$$\frac{4}{3} \frac{\pi b^3}{\langle t \rangle} = \frac{4\pi a D [1 - (a/b)^3]}{\left[(a/b)^{3/2} + 1 \right] [1 - (a/b)^3] - \frac{9}{5} \left(\frac{a}{b} \right) + \frac{3}{2} \left(\frac{a}{b} \right)^3 + \frac{3}{10} \left(\frac{a}{b} \right)^6} \quad (21)$$

When $a/b \to 0$, the Smoluchowski result (Equation 5), which is valid only for infinitely dilute systems, is recovered. An analogous procedure can be used to determine the diffusive-dissociation rate constant.

Two important approximations in Equation 19 should be relaxed if a more quantitative treatment were desired. First, a particle having arrived just inside the diffusive layer (at r_1, where $\delta^* - r_1$ approximates some characteristic dimension of the macromolecule) has a good chance of diffusing out again and being swept away by the flow. When equations such as 17 are used in bulk-solution environments, the usual assumption is that, in equilibrium, conditions outside the shell are on average the same as conditions inside the shell so that for every molecule leaving, another one enters. The boundary is thus treated as a reflector. Our view of the column, however, is that conditions inside the shell where diffusion dominates, are distinctly different from those outside, where transport dominates. Therefore a more accurate approach would be to replace the reflective boundary leading to Equation 17 by a partial sink or a partial source, depending on whether more particles are entering than leaving. In either case, the molecular current at the boundary would be proportional to the concentration there, a complication adding another level of complexity to the theory discussed in subsequent sections. If we confine attention to times neither too early nor too late in the experiment, however, where an approximate steady state is expected to exist locally, the reflective condition might be a reasonably accurate approximation. Clearly, more detailed analyses will be required to evaluate this, but again our philosophy here is only to present a general framework within which the complexities affecting column efficiency can be discussed and to make order of magnitude estimates of the important quantities.

A second important assumption implicit in Equation 18 is that any molecule arriving at the bead surface will be in an appropriate location for reaction. If, however, the surface has only reactive patches, then Equation 19 would have to be modified by a saturable function of the number of patches similar to that in Equation 8.

It is not our intention to pursue all the complexities of this problem here. The main idea to be stressed is that the overall rate for complexation will be some function of each of the three elementary times in Equation 16. For example, if each time is negative exponentially distributed, the result of three in a row would be distributed as a γ-distribution. The diffusive part of this rate is particularly complicated and depends strongly on the width of the diffusive region. This in turn depends on the flow rate. The diffusive part can also depend on any nonspecific potential associated with the bead. Thus, the effective forward-rate constant determined experimentally (see Section D.5) will generally be a complicated function of all these effects, and only under special circumstances will it depend predominantly on the reactive parts k_1 and k_{-1}.

The situation should be contrasted with rates in bulk solution in which convective flow, if it exists, is of a different nature (e.g., stirring), and in which the reactive centers (beads) are generally further apart. In bulk solution too, the overall rate constants will be complicated functions of several processes, but distinctly different from those in a column. Hence, in general, we might not expect rate constants obtained from column chromatography to be the same as those obtained in bulk solution except under conditions for which the reactive step in both cases limits the overall reaction rate. Moreover, both situations are somewhat different than exist in vivo, although we suspect that the column is perhaps a better model than bulk solution for flow-through capillary beds. In principle, a better understanding of the factors entering the expressions for rate constants, will allow their manipulation, and perhaps the achievement of rates that are limited by reaction. This would at least allow us to understand what we are measuring and provide the reactive portions of the rate constants. It is important to note in this regard that recent experimental and theoretical studies[26,27] indicate that for ligand-cell-bound receptor systems, rate constants are far from their diffusion limited values; i.e., they are reaction limited. Since diffusion does not limit the reaction, and it is unlikely that transport in a column limits the reaction, the only question is whether diffusion might be limiting in a column, even if it is not limiting in bulk solution. This difference might arise for example, because a molecule, having entered the Brownian motion domain, can move out again and immediately be swept away by flow. This question needs to be investigated by numerical simulations.

c. Molecular Forces and Their Relation to the Reactive and Diffusive Parts of Rate Constants

The above development makes evident how the forces between the bead and macromolecules influence reaction rates, either with the whole bead or with specific sites on the bead. In this section we complete the discussion by briefly commenting on the physical origin, form, and magnitude of these forces. The implications of these forces for the design of an efficient column are widely recognized (e.g., Lyklema;[28] and Van Oss et al.[29]). Here we will focus on principles and keep the discussion brief by avoiding applications.

At some level, all classical electrostatic forces can be understood in terms of interactions between discrete charges. The simplest and most fundamental form of this interaction is Coulomb's law. If two charges of q_1 and q_2 statcoulombs are held "r" centimeters apart in a medium with dielectric coefficient K (e.g., K = 78 for water at room temperature and 1 for a vacuum), then the magnitude of the force "f" in dynes between them is

$$f = q_1 q_2/(Kr^2) \equiv -\frac{\partial \phi_1}{\partial r} q_2 = -\frac{\partial \phi_2}{\partial r} q_1 \tag{22}$$

$\dfrac{\partial \phi i}{\partial r}$ being the field of charge "i". The force is either attractive or repulsive according to whether the charges are of opposite or similar sign, and it is directed along the line that joins them. Suppose the two charges are of the same magnitude "q", and of opposite sign, and that the line joining them is of length ℓ. An application of the vector form of Coulomb's law indicates that the potential at the end of a vector "r" making an angle θ with ℓ is

$$\phi = \frac{1}{k} \frac{p\cos\theta}{r^2} \tag{23}$$

where $p \equiv q\ell$ and $\ell \gg r$. If the dipole "p" is itself in a field, it will have an energy U given by

$$U = pE\cos\beta \tag{24}$$

where β is the angle between the dipole and the field, E; the latter being the gradient of the potential. All other interaction energies of interest: those due to the various types of van der Waals forces, hydrogen bond forces, and hydrophobic forces, can be viewed as derivative of the dipole interaction, although they cannot all be completely understood without recourse to quantum mechanics.

In order to discuss the types of interactions that arise in a column, we must begin by reviewing the theory for small molecules. The interaction between two polar molecules (e.g., water molecules) separated by a distance that is large compared to their molecular dimensions will be governed by an equation in the form of 24. However, if the dipoles are free to rotate, they will assume a very large number of orientations due to thermal agitation and the forces between them must therefore be appropriately averaged. When the thermal energy is large compared to the dipole-dipole energies, the resulting energy of interaction between two freely orientable dipoles p_1 and p_2 is

$$U = -2p_1^2 p_2^2 / 3K \ kTr^6 \tag{25}$$

where kT is the thermal energy (Boltzmann constant times the temperature).

Attraction is not limited only to molecules with permanent dipole moments. Two other types of van der Waals interactions are possible, arising from the fact that an electric field can induce a dipole in a nonpolar molecule which is proportional to the polarizability α of the molecule. In particular, the interaction energy between a permanent dipole and a dipole that it induces is

$$U = -2p^2\alpha / K^2 r^6 \tag{26}$$

Even when neither type of molecule has a permanent dipole moment, a force of attraction is present. That is because molecules have transient dipole moments which, although they average out over time when the molecules are isolated, do not average to zero when they are close enough to be effected by dipole fluctuations in neighboring molecules. The interaction energy due to this effect is of the form

$$U \cong \frac{3h\alpha_1\alpha_2\omega_1\omega_2}{2n^4 r^6 (\omega_1 + \omega_2)} \tag{27}$$

where "n" is the index of refraction of the medium and "hw" is the ionization potential of the electron giving rise to the induced dipole. As an indication of the relative importance of these forces, we note that with a dielectric coefficient of 1, they are in the ratio of 1:1,100:20,000 for carbon dioxide (apolar) and 1:0.05:0.25 for water (London[30]). Not only do the dispersion forces predominate, as they must for apolar molecules, but they are not entirely insignificant even for polar molecules. The importance of this latter fact is emphasized by recalling that all these forces are important only at very short distances because of their r^{-6} dependence, and it is precisely at such distances (several Angstroms) that dielectric constants, even in polar solvents, are best approximated by values close to one, since very few solvent molecules can modify the field by direct intervention.

The above results indicate that the attractive interactions between two atoms or diatomic molecules are of the form $U = -B/r^6$, where $B \cong 10^{-58}$ erg-cm^6. For large molecules, with additive forces (i.e., all induced dipole fluctuations are assumed to be in phase), the number of terms contributing to the overall interaction is proportional to the square of the number of interacting units and dependent on their spatial distribution. Although the relative importance of each of the three types of forces remains nearly the same as it is for individual atoms, several important differences emerge. First distance falls off much more slowly than r^{-6}. For example, for two large interacting spheres, the potential falls off inversely as the square of the distance.[31] The so called "Hamaker proportionality constant" depends on the types of chemical groups comprising the spheres and on the nature of the intervening medium. When the distance between surfaces is small compared to the sum of the radii, the potential falls off as approximately $1/r$; i.e., the same way the Coulomb potential falls, only with a constant that depends on the polarity of atoms composing the spheres. The assumption of a dielectric constant of order unity is no longer likely to be valid since the separation between the surfaces can be large enough to include many water molecules and yet be small compared to the sum of the radii. The effect of an increased dielectric constant is to make permanent dipole interactions (the first type of van der Waals interaction) relatively more favorable. Finally the assumption that induced dipole fluctuations are in phase breaks down if the molecules are too large since electromagnetic radiation is not able to propagate across the large molecule in a time that is short compared to the period of oscillation of the dipole. The effect becomes important for dimensions that are significantly greater than 100Å, so that for such sizes induced dipole interactions will become relatively less favorable than permanent dipole interactions.

There are two other types of potentials that are of considerable importance, not only for column design, but in the actual folding of proteins. These are due to hydrogen bonds and to hydrophobic interactions; both are poorly understood.

When hydrogen forms a covalent bond with a highly electronegative atom such as oxygen or nitrogen, the spatial probability density of its lone electron is shifted toward the heavy atom. An electron or negative pole of a dipole of another atom or molecule can therefore approach the hydrogen proton closely, though the approach must be highly directional, with the three atoms nearly colinear. The resulting Coulombic interaction is at least in part responsible for the hydrogen bond energy, which is typically in the range of 3 to 7 kcal.

The so called hydrophobic interactions are closely related to van der Walls interactions. An apolar molecule placed in a polar solvent will cause reorganization of the solvent in the vicinity of the molecule, accompanied under certain conditions by an unfavorable free energy which is proportional to some effective area of the apolar molecule. The change arises in a poorly understood way from a distribution of the hydrogen bond energies between water molecules, and substitution of van der Waals interactions. If the apolar molecules are well separated, the total free-energy change will be the sum of the free-energy changes associated with the individual molecules. If, however, the apolar molecules aggregate, their total surface area decreases and hence the total free-energy change is less unfavorable. In addition depending on the size and geometry of the apolar molecules the possibility exists that the interaction between them is more favorable than their interaction with water. This tendency of apolar molecules in a polar environment to aggregate is referred to as the hydrophobic effect, and plays a major role in the design of certain types of columns.[28,29]

B. The Physical Chemical Theory of Liquid Column Chromatography

The use of column chromatography as a quantitative tool for molecular weight determination[3] and chemical reaction characterization[2,5] has increased continuously and rapidly during the past decade. However, the basis of its validity for quantitative thermodynamic (and perhaps kinetic[32]) studies is uncertain, resting largely on assumptions that local chemical equilibrium is established instantaneously[5] or that the contribution of chemical kinetics to elution profile broadening can be made to dominate the effects of diffusion and other nonequilibrium processes.[32] A general assessment of the range of validity of these assumptions has been difficult because of the formidable problems in obtaining analytic solutions for the elution profile even when the mathematical system is linear.

In this section we formulate a model for liquid-column chromatography which is an initial boundary value problem for a system of linear partial differential equations. Explicit solutions involving modified Bessel functions are obtained for the model when diffusion is neglected. Formulas for the mean and variance of the elution profile are then found from these explicit solutions. Explicit solutions are also found when diffusion is not negligible, but these solutions are complicated convolution integrals with a Gaussian Kernal.

Of the many important mathematical contributions to the theory of liquid-column chromatography (see References 2, 33, and the references cited therein), those most similar to the model presented here are the nonequilibrium random-walk model of Giddings and Eyring[33,34] and the partial differential equations formulation of Thomas.[35] Our approach involving explicit equations for the moments of the profile is similar to that used by other authors (see Reference 33 and references cited therein).

Moments of the elution profile can also be found by using a passage-time approach as shown in Sections C and D. This passage-time approach is particularly important because it generalizes to affinity chromatography. Thus, it is possible to find formulas for the mean and variance of the elution profile for affinity-chromatography models even though explicit solutions for the elution profile cannot be found. Hence the theory developed in this section and in Sections C and D provides a foundation for generalization to affinity chromatography.

A chromatography model which assumes local equilibration is described in Section B.5. When diffusion is negligible, this model yields a widely used formula for the peak of the elution profile. Identification of the assumptions required for its derivation provides a basis for deciding when the peak is a reasonable approximation to the mean of the elution profile.

Numerical simulations of the elution profile indicate that the peak and mean may differ by as much as a factor of two for slow mass-transfer into the beads. Since the mean is uniquely determined by the sorption-desorption equilibrium constant, but the peak is not, the use of the peak to characterize the equilibrium constant for broad asymmetric profiles can lead to serious errors. For some low transfer rates the profile has two peaks. This happens even for homogeneous molecules under ideal conditions and is caused by molecules that traverse the bed without penetrating the beads. The dispersion in the first peak is determined by effects other than mass transfer.

1. The Mathematical Model

The literature abounds with detailed descriptions of the experimental procedure.[1,2] Briefly, we will be considering a cylindrical column packed with beads that are composed of cross-linked polymers. On a molecular scale, the interior of a bead can be thought of as a network of tortuous channels of various sizes.

At time $t = 0$, the molecule to be studied is introduced at the top of the column, with the long (or major) axis of the column oriented along the direction of some external field (usually gravitational). Molecules move through the column by transport and diffusion. Movement down the column is, however, delayed for various periods of time as a consequence of molecules diffusing out of the mobile phase (i.e., the void volume between the beads) and into the stationary phase (i.e., the penetrable or interior volume of the beads to which the molecule has access). Since the rates for entering and leaving a bead will depend upon, among other things, molecular size and geometry, the time taken to traverse the column will also depend upon molecular size and geometry.

We model only nongradient, elution chromatography with a constant temperature in the column and a constant pressure on the solvent. The equations describing the elution profile will depend on whether the thickness of the initial solute layer is or is not negligible compared to "h"; i.e., whether a small- or large-zone experiment is being performed. Most of the expressions we derive will describe the former, but some results for the large zone experiments will also be noted and the procedures for obtaining them will be indicated. We first consider experiments with a homogeneous sample, i.e., consisting of a single type of molecule.

Let "x" measure the distance up from the bottom of the bed so that $x = 0$ is the bottom of the bed and $x = h$ is the top. Let V_o be the void volume, i.e, the volume exterior to the beads, and let V_p be the volume interior to the beads that can be penetrated by the molecules. Define the void cross-sectional area as $A_o = V_o/h$ and the penetrable cross-sectional area as $A_p = V_p/h$. The concentration (over A_o) of solute molecules in the mobile phase is $C(x,t)$ and the concentration (over A_p) of molecules in the stationary phase is $B(x,t)$. Thus CA_o and BA_p are the number of molecules per unit distance in each phase. Since we are discussing conditions under which the molecular concentration in any local region of the column can be considered relatively dilute, only a very small fraction of the bead interior will be occupied. Consequently the general nonlinear problem becomes essentially linear, i.e., terms higher than first order in concentration do not appear in the equations. A method for introducing nonlinear effects is presented in Section C.5.

The diffusion-reaction-transport equations are derived using conservation of mass. They are

$$\frac{\partial}{\partial t}(CA_o) = D\frac{\partial^2}{\partial x^2}(CA_o) + u\frac{\partial}{\partial x}(CA_o) - k_1 CA_o + k_{-1} BA_p \tag{28}$$

$$\frac{\partial}{\partial t}(BA_p) = k_1 CA_o - k_{-1} BA_p \tag{29}$$

where k_1 and k_{-1} are the rate constants for transition between the two phases, "u" is the transport velocity, and D is the diffusion coefficient of the molecules along the major axis of the column. The mobile phase flow velocity "u" is equal to F/A_o where F is the flow rate in the column.

Equation 28 describes the rate of change of the mobile-phase concentration in terms of contributions from diffusion (the first term on the right), transport (the second term on the right), and interchanges between the two phases (the last two terms). Equation 29 simply says that when a molecule enters the stationary phase, its average residence time is $1/k_{-1}$; the details of what it may be doing inside the bead being on average unimportant to its large-scale movement. The residence time is, of course, a function of the size and geometry of the molecule. In general the sorption and desorption rate constants k_1 and k_{-1} are concentration dependent, but under the ideal conditions indicated above (dilute molecular concentration), they can be considered essentially constant. Equations 28 and 29 also neglect the possibility of bead-size heterogeneity and nonuniform packing. Both these restrictions can be relieved somewhat (Section C.6).

It is convenient to write Equations 28 and 29 in terms of density functions, i.e., probabilities per unit length of finding a molecule at a particular position in the column. This is accomplished by dividing the equations by I, the total number of molecules initially present. Thus with

$$p \equiv CA_o/I \tag{30}$$

and

$$q \equiv BA_p/I \tag{31}$$

Equations 28 and 29 become

$$\frac{\partial p}{\partial t} = u \frac{\partial p}{\partial x} + D \frac{\partial^2 p}{\partial x^2} - k_1 p + k_{-1} q \tag{32}$$

$$\frac{\partial q}{\partial t} = k_1 p - k_{-1} q \tag{33}$$

At sorption-desorption equilibrium the concentrations C and B are equal so that from Equation 33

$$K = k_1/k_{-1} = q/p = A_p/A_o$$

or, multiplying numerator and denominator by "h"

$$K = V_p/V_o \tag{34}$$

The equilibrium (association) constant K should be distinguished from other coefficients that are also widely used in the literature. In particular, the partition or distribution coefficient is defined as $K_d = V_p/V_i$ where V_i is the volume inside the beads that is not gel matrix and is related to the equilibrium constant K by $K_d = KV_o/V_i$. The retention ratio R used by Giddings[33] satisfies $R = 1/(1 + K)$.

In order to solve Equations 32 and 33 the initial conditions of the experiment must be specified, as must the conditions at the boundaries of the column. With respect to the former, we will assume for the present that the layer at the top of the bed containing sample molecules is small enough to be considered an instantaneous source (small-zone experiment). Thus, the initial conditions on $0 \leq x \leq h$ are

$$p(x,0) = \delta(x - h)$$
$$q(x,0) = 0 \qquad x \neq h \tag{35}$$

The first condition simply says that when solute molecules are first introduced on the column, (t = O) their concentration is zero everywhere except at an infinitesimally thin layer at the top of the bed. This observation is expressed mathematically by the Dirac delta function δ (x − h). The second condition expresses the fact that initially (t = 0) no molecule is interior to the beads. Alternatively the instantaneous source could have been included as a term δ (x − h) δ (t) in Equation 32.

The specification of boundary conditions will, as we explain later, depend on what assumptions we make about the role of axial diffusion (i.e., diffusion along the major axis of the column). For the moment, we will consider the system 32 and 33 without diffusion (D = 0). We can then specify the condition at the top of the column as

$$p(h,t) = 0 \qquad t > 0 \tag{36}$$

Equation 36 says that except at the initial instant of time (t = 0), there is no solute at the very top of the column, a condition that can be strictly valid only when axial diffusion is completely negligible compared to axial transport.

2. The Elution Profile When Transport Dominates Diffusion

For the purpose of finding a solution, we assume that there is no solute buildup at the bottom of the bed, i.e., we assume flow continues as though the bed were extended below x = 0. We thus take the eluted current (molecules per time) at the bottom of a bed of height "h" as

$$\frac{\text{Molecules/sec leaving}}{\text{the bottom of the bed}} = uA_0C(0,t) = uIp(0,t)$$

In this equation I is the total number of solute molecules and $p(x,t)$ is the solution of the initial-boundary value problem with a uniform bed on the semi-infinite interval $-\infty < x \leqslant$ h, with "p" approaching zero as "x" approaches $-\infty$.

Now define $\Delta = ut + x - h$ and $\rho = (k_1 k_{-1}/u^2)\,(h - x)$; then the solution of Equations 32, 33, 35, and 36 found by using Laplace Transforms[36] is

$$p(x,t) = \exp[-k_1(h - x)/u][\delta(\Delta) + H(\Delta)\exp(-k_{-1}\Delta/u)\,\sqrt{\rho/\Delta}\,I_1(2\sqrt{\rho\Delta})] \tag{37}$$

$$q(x,t) = (k_1/u)\exp[-k_1(h - x)/u]\,H(\Delta)\exp(-k_{-1}\Delta/u)\,I_0(2\sqrt{\rho\Delta}) \tag{38}$$

where the Heaviside function H (Δ) is 0 for $\Delta < 0$ and 1 for $\Delta \geqslant 0$, and the symbols I_0 and I_1 are modified Bessel functions.[37] Thomas[35] found solutions similar to 37 and 38 and also found asymptotic approximations. Giddings and Eyring[34] and Giddings[33] obtained similar solutions using a stochastic (random walk) approach.

3. Profile Mean and Variance When Transport Dominates Diffusion

In practice, the complete distributions given by equations 37 and 38 are often not needed. Instead, the mean time for molecules to elute from the column (the first moment of the distribution) and the variance in the elution profile (the second moment about the mean of the distribution) provide the information that an experimentalist needs. Expressions for these quantities can be obtained from Equation 37 by multiplying the molecular current uIp(O,t) by appropriate powers of time (first power if the mean is desired, etc) and integrating over time. In this way we find that the mean elution time is

$$M_e = (1 + K)\,h/u \tag{39}$$

and the variance is

$$S_e = 2Kh/(k_{-1}\,u) \tag{40}$$

These results can be converted to expressions involving elution volumes by multiplying 39 and 40 by the flow rate, F, and its square, respectively (where F, the solute volume per second crossing a plane perpendicular to the column axis equals V_0 u/h). The mean elution volume V_e is defined to be the total volume of solvent eluted up to the mean elution time. We find using Equation 34 that the mean and variance for the eluted volume are given by

$$V_e = V_o + V_p \tag{41}$$

$$W_e = 2FV_oK/k_{-1} = 2FV_p/k_{-1} \tag{42}$$

In Sections C and D we will show that if an expression for the entire profile is not needed or cannot be obtained, methods that are more direct than solving partial differential equations can be used for finding expressions for profile moments.

Solutions for a large solute zone, i.e., for a molecular layer thick enough so that it cannot be considered an instantaneous source, can also be obtained. Although the results are somewhat more complex than those given by Equations 37 and 38, they provide a useful check as an alternative method for arriving at expressions for elution profile characteristics. Details can be found in Section C.4.

4. The Elution Profile When Movement Along the Column by Diffusion Cannot be Neglected

The procedure for finding a solution that includes diffusion along the column can be understood best by considering a molecular interpretation of Equations 37 and 38. In particular, we note that p(x,t) could have been obtained by finding the probability that a molecule at (x,t) had moved freely for a total time $\tau \leq t$, multiplying that probability by the conditional probability that a molecule moving for time τ will be at "x", and then integrating over τ.[33,34] In the absence of diffusion, the kernel in the integrand, i.e., the probability that a molecule moving freely for total time τ is at "x", is $\delta(\tau - (h-x)/u)$ since motion by convection is completely deterministic. This notation simply reflects the fact that all molecules whose time of free movement (i.e., time outside beads) totals to τ, will be located in a thin band at a distance $h-x = u\tau$ from their starting position (i.e., the top of the bed). This procedure uncouples reaction from movement down the bed; something that can always be achieved for a linear system.

If now, rather than allowing movement only by convection, we include diffusion, then the kernel will be Gaussian, rather than a delta function, i.e., the thin band will be subject to diffusive spreading. Thus the delta function will be replaced by

$$G(x,\tau) = (4\pi D\tau)^{-1/2} \exp[-(u\tau + x - h)^2/4D\tau] \tag{43}$$

where $G(x,\tau)$ is the probability that a molecule having moved freely for $\tau \leq t$, will be at "x". The diffusion constant D should be interpreted broadly so that it not only includes the simple Brownian diffusion one would find in a homogeneous medium, but also includes eddy diffusion and velocity profile effects as indicated in Section A.

To find the probability density P(x,t) for free molecules at (x,t), 43 must be multiplied by the reactive probability, i.e., the probability that a free molecule at (x,t) moved freely for $\tau \leq t$. The product must then be integrated over τ from 0 to "t". The probability density Q(x,t) for bound molecules at (x,t) is found similarly. But we already know the solutions for the reactive probabilities since Equations 37 and 38 are simply those solutions integrated over a δ-function kernel. Hence we find that the solutions to the model with diffusion are

$$P(x,t) = \int_o^t G(x,\tau)\, p(h - u\tau,t)\, ud\tau \tag{44}$$

$$Q(x,t) = \int_o^t G(x,\tau)\, q(h - u\tau,t)\, ud\tau \tag{45}$$

These solutions, P(x,t) and Q(x,t), are written out more explicitly in Reference 8. The functions P and Q above satisfy the differential equations 32 and 33, the initial conditions 35, and the boundary conditions $P(\pm \infty,t) = 0$, $Q(\pm \infty,t) = 0$.

The main approximation in 44 and 45, for which we expect the error to be negligible, is that they hold on the interval $-\infty < x < \infty$, whereas the chromatographic bed only occupies $0 \leqslant x < h$. The contribution from molecules that move above "h" is expected to be only a second order effect since only a small fraction of solute will ever be above "h" when movement down the bed is dominated by convection. We shall, however, consider the magnitude of the error in Section D.4. Similarly we expect the concentration of molecules just above $x = 0$, with no bed beneath $x = 0$, to be essentially the same as it would if the bed continued below $x = 0$.

The numerical solution to these equations, although possible, is not the most useful approach to obtaining quantitative estimates of the key parameters controlling the form and dispersion of the elution profile. In fact if one is content with only certain characteristics of the profile — its mean, variance, skewness, and so forth — relatively simple expressions can be obtained by methods to be outlined later in Sections C and D.

Approximate solutions to the large zone problem with diffusion can be found as above by convoluting the large zone solution without diffusion with the Gaussian kernel Equation 43. Before discussing the very powerful techniques for obtaining exact expressions for profile moments, we briefly review approximate expressions for elution profiles.

5. Approximate Expressions for Elution Profiles
a. Local Equilibration and the Profile Peak

The widely used equation for the elution profile peak can be obtained by assuming that (outside the boundary layer, as explained in Section B.2) transport dominates axial diffusion but that local equilibrium between mobile and immobile phase molecules nevertheless prevails. Because of this latter assumption, at each position "x", $B(x,t) = C(x,t)$ so that $q(x,t) = Kp(x,t)$. The sum of Equations 32 and 33 then reduces to

$$\frac{\partial p}{\partial t} = \frac{u}{1 + K}\frac{\partial p}{\partial x} + \frac{D}{1 + K}\frac{\partial^2 p}{\partial x^2} \tag{46}$$

The solution of the initial value problem for $-\infty < x < \infty$ consisting of 46 and $p(x,0) = \delta(x-h)$ is

$$p(x,t) = [4\pi Dt/(1 + K)]^{-1/2} \exp\left[\frac{-(h - x - ut/(1 + K))^2}{4Dt/(1 + K)}\right] \tag{47}$$

This solution has the form of a Gaussian distribution function that moves and spreads simultaneously. The elution profile is given by

$$\text{Particles per second crossing the bottom of the bed} = up(0,t) + D\frac{\partial p}{\partial x}(0,t)$$

The profile is not a Gaussian distribution as a function of time and it is not symmetric around the peak which occurs at

$$T(\text{peak}) = \frac{(1 + K)h}{u}\left(1 - \frac{7}{3r} + \frac{91}{54r^2} + \cdots\right) \tag{48}$$

where $r \equiv uh/D$. Evidently if "r" is very much greater than 1, transport totally dominates diffusion, the elution profile is approximately $\delta[(h - x - ut)/(1 + K)]$, and the peak approximates the mean. Hence the corresponding formula

$$V(peak) = V_o + V_p \qquad (49)$$

assumes both local equilibration and negligible diffusion while the formula

$$V_e = V_o + V_p \qquad (50)$$

for the mean elution volume assumes only that diffusion is negligible. The distinction between the peak and the mean is the subject of the next section.

b. Comparison of the Peak and Mean Eluted Volumes

As indicated above, mean volumes are often more convenient to use than mean times. Thus, the transport dominant result for the volume eluted during the mean time required by a molecule to traverse the column is

$$V_e = V_o + V_p \qquad (41)$$

For simplicity we will refer to V_e, as the mean eluted volume; it is to be carefully distinguished from the peak eluted volume.

We emphasize that Equation 41 was derived without equilibrium assumptions. The result is generally valid under nonequilibrium conditions subject only to the assumptions underlying Equations 37 and 38 as a model for column chromatography and the recognition that the contribution of diffusion to the mean value of the elution profile is small compared to the contribution of transport.

Equation 41 is formally identical to Equation 49 for the peak of the elution profile, the latter being a standard equation in the theory of molecular sieving; however, the interpretations of these equations differ. The standard equation says that the peak-eluted volume varies linearly with the penetrable volume; Equation 41 says that the mean-eluted volume varies linearly with the penetrable volume. The mean is of course a weighted volumetric average over the entire profile and, in general, is equal to the peak only for symmetric profiles. For the model represented by Equations 32 and 33, the peak and mean are exactly equal only when the initial layer moves down the column as a sharp front. This happens under local equilibrium conditions, but then there is no dispersion in the profile due to nonequilibration. The assumption of local equilibration in the sorption-desorption kinetics is usually unreasonable since it implies that the variance is approximately zero. It is perhaps worth noting at this point, that without the concept of diffusive boundary layer, introduced in Section A, the assumptions that transport dominates diffusion and that local equilibrium prevails — both of which are required for the validity of Equation 49 — are unlikely to hold simultaneously since equilibrium between the void and penetrable volumes can only be established by diffusion, and local equilibration requires that the time for these diffusive processes must be fast compared to changes caused by transport.

Aside from the weak conceptual basis for Equation 49, its practical efficacy will in many cases be suspect. According to Equation 41, with the void volume fixed, the penetrable volume is uniquely determined by the mean-eluted volume. This is not true for the peak-eluted volume which in general depends on V_p and either k_1 or k_{-1}.[33] The distinction is important when molecular weight is low ($\sim 40,000$) as mass determination may then be in error by as much as a factor of two. Moreover — and this will be especially important for bivalent antibodies — Equation 41 holds under nonequilibrium conditions, whereas the expression for V(peak) assumes local equilibrium conditions. For large k_1 and hence small dispersion, the profile peak and mean are approximately the same. As k_1 decreases, the peak becomes less than the mean. Graphical comparisons of the peak and mean of the elution profile as a function of k_1 are given in Reference 8.

C. The Moments of the Elution Profile

As we discussed in Section B, the initial-boundary value problem need not be solved explicitly if our primary interest is in the first few moments (mean, variance, and skewness) rather than the entire profile. In Section B the mean and variance of the elution profile were found from explicit solutions for the elution profile. The moments of the elution profile can be found by solving an ordinary differential equation derived from the partial differential equations in the chromatography model. In this section we will show how these expressions can be obtained exactly for column models and boundary conditions that are somewhat more realistic than those used in Section B. We note at the outset, however, that these more realistic, and hence more complex, expressions need not always be used in place of the simpler expressions. Their importance lies in completeness; they provide the relations required to assess the conditions under which the simpler expressions are and are not valid. Both small- and large-zone chromatography models are considered. Detailed derivations for a very general model are given in Section D.

A chromatography model with nonlinear sorption-desorption kinetics is considered in Section C.5. When the mass transfer rates are concentration dependent, the sorption-desorption equilibrium constant is no longer uniquely determined by the profile mean. However, uniqueness is obtained in the limit as both mass-transfer coefficients become small, with their ratio remaining moderate. Graphs reveal the remarkable dependence of the mean of the elution profile for the nonlinear model on the mass-transfer rates.

In Section C.6 we consider heterogeneity or nonuniformity in the bed and in the molecules. When equilibrium constants are distributed, the expressions for the mean time involve the average equilibrium constant (or the average penetrable volume). If the sorption rate is constant, then the variance of the elution profile is proportional to the sum of the square of the average equilibrium constant and the variance of the distribution.

1. Definitions

We begin by defining the mean passage time $T_1(x)$ as the mean time[38] for molecules starting in the initial layer to move past a position "x". The mean elution time $M_e = T_1(0)$ (also called the mean residence time or the mean retention time) is the mean time for molecules to move out of the bottom of the bed. Definitions of higher moments are similar. The fraction of molecules per unit time moving by position "x" in the bed at time "t" is the current $up(x,t) + D\dfrac{\partial p}{\partial x}(x,t)$. Thus, the "jth" moment of the time to reach position "x" is defined as

$$T_j(x) = \int_o^\infty t^j \left[up(x,t) + D\,\frac{\partial p}{\partial x}(x,t) \right] dt \tag{51}$$

2. Mean Time for Passage Through the Column
a. Mean Elution Time When Axial Diffusion is Negligible

We are interested in the mean time required for the average molecule to reach a position "x" in the column and, more specifically, in the mean time for elution from a column packed to a height "h".

We will first suppose that both the bed and the solvent extend only to a height "h". In experimental preparations, solvent extends above "h" and this complicates models which include axial diffusion since solute can diffuse above "h". We expect this effect to be small, but analyze it fully in Section D.2.

Starting with the definition

$$T_1(x) = \int_o^\infty tup(x,t) \, dt \qquad (52)$$

it is reasonably easy to show directly (cf. Section D.2) that T_1 satisfies the differential equation

$$\frac{dT_1}{dx} = -\frac{(1 + K)}{u} \qquad (53)$$

where $K = k_1/k_{-1}$. See Reference 39 for an alternate definition of the mean passage time and a derivation of Equation 53. Since by definition, particles that start at $x = h$ take no time to arrive there, Equation 53 is solved subject to $T(h) = 0$. Thus

$$T_1(x) = (1 + K)(h - x)/u \qquad (54)$$

Therefore the mean time for arrival at the bottom of the column is

$$T_1(0) \equiv M_e = (1 + K)h/u \qquad (39)$$

which was derived in a much more complicated manner in Section B.3.

Equation 39 assumes that the equilibrium distribution between the void and penetrable volumes is determined only entropically, i.e., nonspecific interaction between molecule and bead is assumed to be negligible. In the presence of nonspecific binding a multiplicative constant would precede V_p.

b. Mean Elution Time When Axial Diffusion is Significant: An Approximate Treatment

When diffusion along the length of the column is included, the derivation is more complicated, but the result changes in a numerically trivial manner. Suppose the bed and solvent both extend to height "h". Then with $r = uh/D$

$$T_1(0) = (1 + K)(h/u)\{1 - [1 - \exp(-r)]/r\}$$

or, in terms of volume

$$V_e = (V_o + V_p)\{1 - [1 - \exp(-r)]/r\} \qquad (55)$$

A complete analysis will be presented in section D.2. Here we note that if $1/r \ll 1$, diffusion is unimportant. For example, if $D = 10^{-6}$ cm²/sec, h = 25 mm, and u = 0.04 mm/sec, then $r = 10^{-4} \ll 1$. More generally if diffusion is to contribute less than 1% to the mean time (clearly less than experimental error), a reasonable criterion for neglect of diffusion is

$$1/r < 0.01$$

which implies that

$$r > 100 \qquad (56)$$

3. Higher Order Moments
a. Dispersion in the Elution Profile

The fact that the thin layer of molecules at the top of the column is not eluted as a thin layer can be attributed in large part to nonequilibrium effects and perhaps also to heterogeneity in bead packing and size, and in the heterogenity of the sample molecules. Heterogeneity is not considered until Section C.6. In Section C.2 we developed an expression for the mean of the elution profile. In this section, we comment on the dispersion in the elution profile.

If we define the profile variance as

$$S(x) = T_2(x) - T_1(x)^2 \tag{57}$$

then the differential equation for the dispersion about the mean, with transport dominating diffusion, is

$$u \frac{dS}{dx} = - \frac{2K}{k_{-1}} \tag{58}$$

from which we find using $S(h) = 0$ that

$$S_e = S(0) = \frac{2}{k_{-1}} \frac{h}{u} K \tag{59}$$

or, in terms of volume eluted, the variance is

$$W_e = F^2 S_e = 2FV_o K/k_{-1} = 2FV_p/k_{-1} \tag{60}$$

b. Profile Skewness

The above methods can be readily continued for higher moments. In particular we find that (neglecting diffusion) the third central moment (i.e., around the mean) of the elution profile as a function of time is

$$N_e = 6 Kh/(k_{-1}^2 u) \tag{61}$$

and as a function of volume is

$$U_e = 6F^2 V_o K/k_{-1}^2 \tag{62}$$

When the flow rate is such that diffusion is negligible, the moment coefficient of skewness for a thin sample layer is

$$G_1 = U_e/W_e^{3/2} = 3(F/2k_1 V_o)^{1/2} \tag{63}$$

Thus, the skewness increases as the flow rate F increases as noted in numerical solutions.[40]

4. Profile Mean and Dispersion for Large Zone Experiments

If the effects of diffusion can be neglected and T is the thickness of the initial layer of sample, then the means and variances given by Equations 39, 41, 59, and 60 become

$$M_e = (1 + K) h/u + T/2u \qquad \text{mean-elution time}$$

$$S_e = 2Kh/(k_{-1}u) + T^2/(12u^2) \qquad \text{variance in elution time} \tag{64}$$

$$V_e = V_o + V_p + V_oT/2h \qquad \text{mean-eluted volume}$$

$$W_e = 2FV_oK/k_{-1} + V_o^2T^2/(12h^2) \qquad \text{variance around mean-eluted volume} \qquad (65)$$

Since the third central moment of the time for a layer of thickness T to enter the bead is zero, Equations 61 and 62 hold for both thin and thick sample layers. See Reference 7 for an explicit solution for a large solute zone experiment.

5. Nonlinear Kinetics

Consider the chromatography model without diffusion for the thin solute zone described in Section B. Assume that the sorption-desorption kinetic rates decrease near saturation, i.e., the kinetics are no longer linear. Then Equations 32 and 33 become

$$\frac{\partial p}{\partial t} = u\frac{\partial p}{\partial x} - k_1 p(1 - C/C^m) + k_{-1}q(1 - B/B^m) \qquad (66)$$

$$\frac{\partial q}{\partial t} = k_1 p(1 - C/C^m) - k_{-1}q(1 - B/B^m) \qquad (67)$$

where the parenthesized factors reflect saturation effects. The mobile and stationary phase concentrations, C and B, are less than the corresponding maximum concentrations, C^m and B^m. Since $p = CA_o/I$ and $q = BA_p/I$, we find

$$C/C^m = ph(I/C^mA_oh) \qquad (68)$$

$$B/B^m = qh(I/B^mA_ph) \qquad (69)$$

where A_oh and A_ph are the void and penetrable volumes, respectively. If the total number of molecules I is much less than the molecular capacities C^mA_oh and B^mA_ph, then a suitable small parameter is

$$\epsilon = I/C^mA_oh \qquad (70)$$

Since $K = V_p/V_o$, then

$$I/B^mA_ph = \epsilon C^mV_o/B^mV_p = \epsilon a/K \qquad (71)$$

where $a = C^m/B^m$ is a constant near 1 since the saturation concentration C^m and B^m are nearly equal.

Let the expansion of p and q in powers of ϵ be

$$p = p_o + p_1\epsilon + p_2\epsilon^2 + \cdots$$

$$q = q_o + q_1\epsilon + q_2\epsilon^2 + \cdots \qquad (72)$$

After modifying Equations 66 and 67 to include ϵ, substituting Equations 72, and equating like powers of ϵ, we find that p_o and q_o are given by Equations 32 and 33 and p_1 and q_1 satisfy

$$\frac{\partial p_1}{\partial t} = u\frac{\partial p_1}{\partial x} - k_1(p_1 - hp_o^2) + k_{-1}\left(q_1 - \frac{ah}{K}q_o^2\right)$$

$$\frac{\partial q_1}{\partial t} = k_1(p_1 - hp_0^2) - k_{-1}\left(q_1 - \frac{ah}{K}q_0^2\right)$$

$$p_1(x,0) = 0, \quad q_1(x,0) = 0, \quad p_1(h,t) = 0 \tag{73}$$

The first-order approximation to the mean elution time is

$$M_e = \int_0^\infty tu[p_o(0,t) + \epsilon p_1(0,t)] \, dt = (1 + K)(h/u) + \epsilon \int_0^\infty tup_1(0,t) \, dt \tag{74}$$

Following the procedure in the section "The Moments of Elution Profile" in Reference 7, we find

$$\int_0^\infty tup_1(x,t) \, dt = -\int_x^h \left[K \int_0^\infty hp_0^2(w,t) \, dt - \frac{a}{K} \int_0^\infty hq_0^2(w,t) \, dt \right] dw \tag{75}$$

Substituting the expressions in Equations 32 and 33 for p_O and q_O in 75 and using identities involving integrals and series,[41] we obtain

$$\int_0^\infty tup_1(0,t) \, dt = -(h/u) \left[\frac{2 + \beta}{4\beta} (1 - e^{-2\alpha}) - \alpha e^{-2\alpha} + \frac{1}{2} \int_0^\alpha e^{-w} I_o(w) \, dw \right.$$

$$\left. \left(\frac{\alpha}{2}\right) e^{-2\alpha} I_1^2(\sqrt{2\alpha}) - \frac{a}{2} \int_0^\alpha e^{-w} I_o(w) \, dw \right] \tag{76}$$

where $\alpha = k_1 h/u$ and $\beta = k_{-1} h/u$.

Numerical evaluation of the integral in Equation 76 indicates that it is well approximated, over a wide range of α, by

$$\int_0^\alpha e^{-w} I_o(w) \, dw \cong -0.1895 + 0.855 \sqrt{\alpha} \tag{77}$$

In particular, this approximation holds to within 6% at $\alpha = 0.2$; 1% at $\alpha = 1$, and it becomes increasingly better as α increases. Therefore Equation 74 can be written as

$$M_e = (1 + K)(h/u) - \epsilon(h/u)\left\{ \frac{2 + \beta}{4\beta} (1 - e^{-2\alpha}) - \alpha e^{-2\alpha} \right.$$

$$\left. - \frac{\alpha}{2} e^{-2\alpha} I_1^2(\sqrt{2\alpha}) - \frac{1}{2} (1 - a)(0.1895 - 0.855) \sqrt{\alpha}) \right\} \tag{78}$$

If α and β are small, with K remaining moderate, the result simplifies to

$$M_e \simeq (1 + K)(h/u) - \frac{\epsilon hK}{u} = [1 + K(1 - \epsilon)](h/u) \tag{79}$$

The relative error in neglecting the nonlinear kinetics can be estimated by combining Equations 74 and 76. Then, to a first approximation

$$\left| \frac{M_e - (1 + K)(h/u)}{(1 + K)(h/u)} \right| \leq \frac{\epsilon}{1 + K} \left[\frac{2 + \beta}{4\beta} + (1 + a) \sqrt{\frac{\alpha}{2\pi}} \right] \tag{80}$$

Equation 78 predicts that in the nonlinear regime, the elution profile mean depends separately on the mass-transfer rates k_1 and k_{-1}, not just on their ratio K. Thus with K held fixed, a change in $\alpha = k_1h/u$ can cause changes in the mean (Figure 1). The result clearly illustrates that when sorption-desorption kinetics are nonlinear, the mean is not uniquely determined by the equilibrium constant K, but depends on the sorption and desorption rate constants k_1 and k_{-1} separately. Note that the linear mean is usually larger than the nonlinear mean. More interesting are the maxima and minima that occur in some simulations. Figure 1 clearly illustrates that when the sorption-desorption kinetics are nonlinear, the mean is not uniquely determined by the equilibrium constant K, but depends on the sorption and desorption rate constants k_1 and k_{-1}. The two peaks reflect two sources of dispersion: sorption-desorption kinetics and axial diffusion.

6. Heterogeneity in the Solute Sample or in the Chromatography Bed

The models presented thus far assume that the molecules are uniform in size, structure, and weight and that the beads (gel particles) are uniform in their packing, size, and structure. Here we examine the effect of nonuniformities on the moments of the passage time. Assume that the equilibrium constant K is distributed with probability density function n(K). Note that the distribution of K could be due to a distribution of k_1, k_{-1}, or both. The average over the distribution of K of the "jth" moment of the passage time at position "x" is defined as

$$\overline{T}_j = \int_o^\infty n(K) \int_o^\infty t^j[up(x,t;K) + D\frac{\partial p}{\partial x}(x,t;K)]\, dt\, dK \tag{81}$$

For a small zone experiment in which diffusion is negligible, the analogs of the expressions 39 through 42 for the moments are

$$\overline{M}_e = (1 + \overline{K})\, h/u \qquad \overline{S}_e = 2\overline{(K/k_{-1})}\, h/u \tag{82}$$

$$\overline{V}_e = V_o + \overline{V}_p \qquad \overline{W}_e = 2FV_o\overline{(K/k_{-1})} \tag{83}$$

where $\overline{K} = \overline{V}_p/V_o$ and $\overline{(K/k_{-1})}$ is the expected value of K/k_{-1}.

If the main sources of equilibrium-constant heterogeneity are nonuniformities in the beads or in their packing, then the average equilibrium constant $\overline{K}$ and the average penetrable volume $\overline{V}_p$ depend only on the column conditions. As long as the same column is used under the same conditions (for example, during molecular weight calibration and determination), this nonuniformity does not significantly affect the results since it is the same for all molecules passing through the column.

If the main sources of equilibrium-constant heterogeneity are nonuniformities in the size, structure, or weight of the molecules, then $\overline{K}$ and $\overline{V}_p$ are averages over these molecules. If the sorption rate k_1 is always the same for the molecules, but the desorption rate k_{-1} varies because of differences in the molecules, then $\overline{(K/k_{-1})} = \overline{K^2}/k_1$. In this case, if k_1 is known (e.g., from another experiment), then the second moment $\overline{K^2}$ can be estimated from the observed variance of the elution profile. Hence the variance $\overline{W}_e$ of the elution profile satisfies

$$\overline{W}_e = \frac{2FV_o\overline{K^2}}{k_1} = \frac{2FV_0}{k_1}\left[(\overline{K})^2 + \overline{[K^2} - (\overline{K})^2]}\right] \tag{84}$$

where the first term in the brackets is the square of the mean of the equilibrium constant

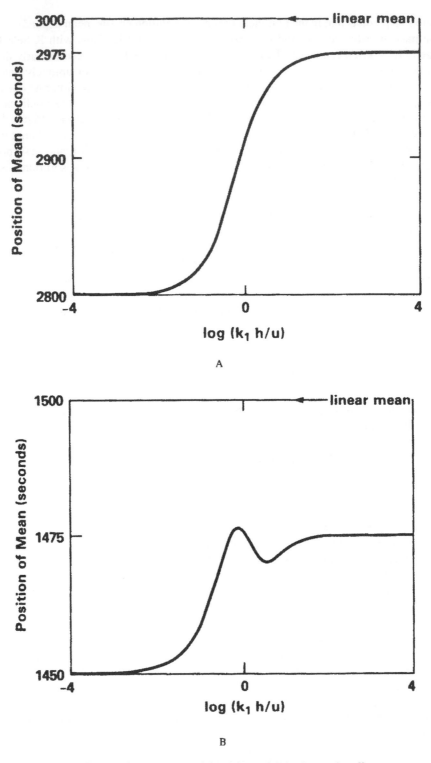

FIGURE 1.   The position of the mean-eluted volume when bead saturation effects are present is not uniquely determined by the sorption-desorption equilibrium constant K. In all panels a = 1 and ϵ = 0.1. (A) K = 2, (B) K = 0.5, (C) K = 0.08, and (D) K = 0.01.

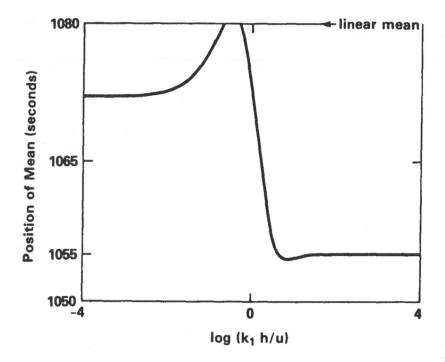

FIGURE 1C

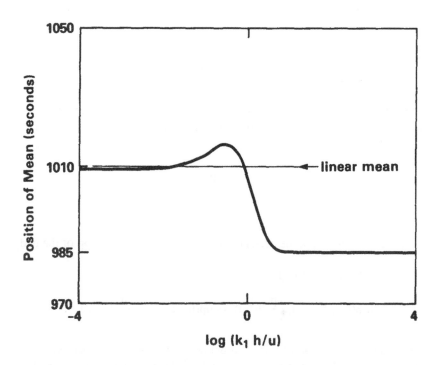

FIGURE 1D

distribution and the second term is the variance. Thus, if k_i is constant, both the means and the variances of the distribution can be estimated.

D. Derivation of Elution Profile Moments When Axial Diffusion is Significant

Here we formulate a model for liquid-column chromatography which includes nonequilibrated mass transfer, diffusion, and very general boundary conditions at the top of the column. A passage-time approach is then used to derive expressions for the moments of the elution profile. The formulas for the mean and variance reduce to those obtained in earlier sections when diffusion is negligible. Precise criteria on the flow rate are given, which guarantee that the effects of diffusion on the means and variances can be neglected. Thus, the very general formulas allow us to determine the range of validity of the simpler formulas given in previous sections.

In Section D.5, a complete procedure is outlined for estimating the equilibrium constant and the rate constants for the sorption-desorption kinetics from the mean and variance of the elution profile. Both the equilibrium constant and the rate constants can be estimated from the mean and variance, but only the equilibrium constant can be estimated from the mean.

1. Modeling the Column Boundaries

For a model including diffusion, molecules can move into the solvent above the bed. We expect this effect on the profile to be small and present here an analysis which confirms this expectation.

The partial differential equation for the mobile-phase molecule density A_1C above the top of the bed ($h < x < f$) involves flow and diffusion (A_1 is the cross-sectional area of the column). We can write a differential equation for A_1C just as we did for A_oC in Section B.1. When this is done and the equation is divided by I, we obtain the following differential equation for the probability density $p(x,t)$

$$\frac{\partial p}{\partial t} = u_1 \frac{\partial p}{\partial x} + D_1 \frac{\partial^2 p}{\partial x^2} \qquad (85)$$

Above the bed the velocity u_1 of the solvent is F/A_1 and the diffusion constant of the molecules is D_1. Since no molecules can move above the top of the solvent, the molecular current there is zero; that is,

$$u_1 p(f,t) + D_1 \frac{\partial p}{\partial x}(f,t) = 0 \qquad (86)$$

for $t \geq 0$.

The number of molecules and the current must be continuous at the top of the bed so that

$$p(h^-,t) = p(h^+,t)$$

$$u_1 p(h^+,t) + D_1 \frac{\partial p}{\partial x}(h^+,t) = up(h^-,t) + D \frac{\partial p}{\partial x}(h^-,t) \qquad (87)$$

for $t > 0$. Precise incorporation of diffusion below the bottom of the bed would require a differential equation like Equation 85 for the solute molecules below the bed and matching conditions similar to Equation 86 at the bottom of the bed. It is more convenient to assume that the elution profile is measured at the bottom of the bed ($x = 0$) and that there is an absorber (sink) there. This corresponds to $A_oC(0,t) = 0$ or

$$p(0,t) = 0 \qquad (88)$$

for $t \geq 0$. This condition is reasonable since the rate of movement by flow of solute molecules in the solution below the bed is large compared to the rate of movement of these molecules in the bed. This model is also applicable to affinity chromatography in which molecules covalently bound to the surfaces of impenetrable beads interact monovalently with solute molecules. In that case the equilibrium and rate constants describe the binding reaction.

2. Mean Passage Time

We now derive an expression for the mean-elution (passage) time, which was defined in Section C.1 as the mean time for molecules to move out of the bottom of the bed. Direct integration with respect to "t" on the interval $(0, \infty)$ of 1 times and "t" times the partial-differential Equations 32, 33, and 85 and use of the boundary conditions in Section D.1 leads to the following boundary value problem for $T_1(x)$.

$$D_1 \frac{d^2 T_1}{dx^2} + u_1 \frac{dT_1}{dx} = 0 \qquad h < x < f$$

$$D \frac{d^2 T_1}{dx^2} + u \frac{dT_1}{dx} = -(1 + K) \qquad 0 < x < h \qquad (89)$$

The boundary conditions 86 and 87 convert to

$$T_1(f) = 0, \qquad \frac{dT_1}{dx}(0) = 0$$

$$T_1(h^+) = T_1(h^-), \qquad (1 + K)\frac{dT_1}{dx}(h^+) = \frac{dT_1}{dx}(h^-) \qquad (90)$$

The solution of this boundary-valued problem is given in Reference 7. The mean-elution time obtained from Equations 89 and 90 is

$$M_e = T_1(0) = \frac{h}{u}\left\{(1 + K)\left[1 - \frac{1 - \exp(-r)}{r}\right] + \frac{1 - \exp(-r)}{r_1}[1 - \exp(r_1 - r_2)]\right\} \qquad (91)$$

where $r = uh/D$, $r_1 = u_1 h/D_1$ and $r_2 = u_1 f/D_1$.

3. Variance of the Passage Time

If the differential Equations 32, 33, and 85 are multiplied by t^2 and integrated with respect to "t" on $(0, \infty)$, then the boundary-value problem for the variance $S(x) = T_2(x) - T_1(x)^2$ is

$$D_1 \, d^2S/dx^2 + u_1 \, dS/dx = -2D_1(dT_1/dx)^2 \qquad h < x < f$$

$$D \, d^2S/dx^2 + u \, dS/dx = -2D(dT_1/dx)^2 - 2K/k_{-1} \qquad 0 < x < h \qquad (92)$$

$$S(f) = 0, \; S(h^+) = S(h^-), \; \frac{dS}{dx}(0) = 0$$

$$\frac{dS}{dx}(h^-) = (1 + K)\frac{dS}{dx}(h^+) + \frac{2K}{k_{-1}}\frac{dM}{dx}(h^+) \qquad (93)$$

From the solution $S(x)$ the variance of the elution time is

$$S_e = S(0) = \frac{2Kh}{k_{-1}u}\left[1 - (1 - \exp(-r))/r\right]$$

$$+ \frac{2(1 + K)^2 h^2}{r\, u^2}\left[1 + 2e^{-r} - 2\frac{(1 - e^{-r})}{r} - \frac{(1 - e^{-2r})}{2r}\right]$$

$$+ \frac{2(1 + K)\, h^2}{r_1 u^2}\left[1 - \exp(r_1 - r_2)\right]\left[\frac{1 - e^{-r}}{r} - 2e^{-r}\right]$$

$$+ \frac{h^2}{r_1^2\, u^2}(1 - e^{-r})^2 (1 - e^{r_1 - r_2})^2 \tag{94}$$

Special cases of some of the equations in this section were obtained earlier.[39] When diffusion is ignored ($D = 0$, $D_1 = 0$), expressions for the mean and variance of the elution profile reduce to equations

$$M_e = (1 + K)\, h/u, \quad S_e = 2Kh/(k_{-1}u) \tag{95}$$

or, in terms of eluted volume

$$V_e = FM_e = V_o + V_p, \quad W_e = FS_e = 2FV_o\, K/k_{-1} \tag{96}$$

Note that the mean-elution volume V_e does not depend on the flow rate F, but the variance increases as the flow rate increases.

4. When is Diffusion Negligible?

From Equation 94 we see that diffusion contributes more to dispersion for low-flow rates and that nonequilibration of mass-transfer contributes more for high-flow rates. For many chromatography systems an intermediate-flow rate is chosen so that the axial dispersion is minimized.[42-44] Here we choose a flow rate (higher than the dispersion-minimizing flow rate) so that the effects of diffusion are negligible. This procedure allows us to use the information contained in the observed dispersion to estimate the sorption- and desorption-rate constants.

The expressions for the mean and variance of the elution time are particularly simple when the contribution of terms corresponding to diffusion are sufficiently small so that they can be neglected. From Equation 91 we find the following bound for large-flow rate on the relative difference between the mean-elution times with and without diffusion

$$\left|\frac{M_e - (1 + K)\, h/u}{(1 + K)\, h/u}\right| \leq \frac{1}{r} + \frac{1}{r_1(1 + K)} \tag{97}$$

For a large-flow rate F, the relative magnitude of the terms corresponding to diffusion in Expression 94 for the variance are given in the following inequality.

$$\left[\frac{S_e - 2Kh/(k_{-1}u)}{2Kh/(k_{-1}u)}\right] \leq \frac{1}{r} + \frac{k_{-1}h}{Ku}\left[\frac{(1 + K)^2}{r} + \frac{(1 + K)}{rr_1} + \frac{1}{2r_1^2}\right] \tag{98}$$

For positive K it is clear that the contributions to the mean-elution time and to the variance of the diffusion-related terms can be neglected if the flow rate (and hence u, u_1, r, and r_1) is chosen large enough so that the right sides of Equations 97 and 98 are small. Note that the above inequalities are also measures of the relative contributions due to diffusion for the mean and variance of the elution volume. Calculations using typical parameter values reveal that the contribution of diffusion above the bed is much less than the contribution of diffusion in the bed and that diffusion terms have a larger influence on the variance than on the mean.

5. The Experimental Procedure

We outline a precise procedure for determining the sorption-desorption equilibrium constant K, the sorption-rate constant k_1 and the desorption-rate constant k_{-1} by liquid column chromatography. This procedure works for both small- and large-zone chromatography.

1. Use Equations 97 and 98 with estimates for the parameter values to choose the flow rate F large enough so that the contributions of diffusion are negligible.
2. Find the void volume V_o experimentally by using the mean-elution volume as V_o when a weaker solvent or a nonretained larger molecule is run through the column.
3. Find values of the mean volume V_e and the variance W_e of the elution profile by running the sample through the column.
4. Calculate the equilibrium constant K from

$$K = V_p/V_o = [V_e - V_o - V_oT/(2h)]/V_o \qquad (99)$$

by using the values determined in Steps 2 and 3. The equations in this step are obtained from Equation 64. Other equations in 64 would be used if the elution profile were given as a function of time. The term in Equation 75 involving the sample thickness T can be omitted if the volume measurement is started when half of the sample has entered the column. Now calculate the desorption-rate constant from

$$k_{-1} = 2F V_oK/[W_e - V_o^2T^2/(12h^2)] \qquad (100)$$

The term involving the sample thickness T is always included but may be negligibly small if T is small. Of course, the sorption-rate constant can now be found from $k_1 = k_{-1} K$.
5. Use the values calculated above in Equations 97 and 98 to verify that the contributions of diffusion were indeed negligible.

We emphasize that the sorption and desorption rates can be found by using the first and second moments, but only the equilibrium constant can be found from the first moment. The third central moment does not yield any new information; however, the observed third central moment can be used in Equation 62 to check the estimates of K and k_{-1} obtained from the first two moments.

If the only constant of interest is the equilibrium constant (and not the rate constants), then the most accurate method of determining this value might be to choose a chromatography system so that the dispersion (band width) is minimized.[42,44] The dispersion can be made quite small by using optimum bed-particle sizes, column diameters and lengths, flow rates, pressures, sample dilutions, and sample sizes in high-pressure liquid chromatography.[44] In this case the mean and peak of the elution profile would essentially coincide so that the equilibrium constant could be easily estimated. We emphasize that if the goal is to also estimate the rate constants for movement in and out of the beads, then the procedure outlined at the beginning of this section should be followed. Estimates of the sorption- and desorption-rate constants are necessary for the quantitative analysis of more complicated systems that include chemical reaction.[9,33]

E. Column-Scanning Chromatography

Column scanning involves measuring the concentration of solute along the bed at various times.[2,38] If the flow rate is such that diffusion can be neglected, then the solute profiles are given by $p(x,t) + q(x,t)$ where "p" and "q" are given by Equations 37 and 38 for a small-zone experiment. There is no explicit expression for the theoretical peak of the solute-concentration profile down the column since it is the root of a transcendental equation obtained by setting the derivative of $p + q$ with respect to "x" equal to zero. Two relatively simple alternatives exist.

If rapid equilibration of the solute-bed kinetics is assumed for a small-zone experiment, then the solute profile is $p + q = (1 + K)p$ where "p" is given by Equation 47. The peak for fixed "t" of the solute profile occurs when $ut = (1 + K)(h - x)$. since $K = V_p/V_o$ and the volume eluted, satisfies $V = V_o ut/h$, the elution volume when the profile peak is at "x" is

$$V = (A_o + A_p)(h - x) \qquad (101)$$

The derivative of this eluted volume with respect to the distance $h - x$ down the bed is

$$\frac{\partial V}{\partial(h - x)} = A_o + A_p = A_o + K_d A_i \qquad (102)$$

Brumbaugh and Ackers[45] plotted the volume at the peak position against the distance down the bed for several molecules and used the observed slopes to estimate the distribution coefficients K_d. Their method has the advantage that more data are obtained from each experiment; however, the rapid-equilibration assumption is likely to find varying degrees of success.

An alternative procedure for which the mathematics can be carried out exactly, and which has the additional advantage of yielding an estimate of k_{-1}, relies on a modified definition of the residence time. Choose the flow rate so that diffusion can be neglected and define the moments of a modified residence time in the bed above position "x" to be

$$R_1(x) = \int_o^\infty t[p(x,t) + q(x,t)] \, dt / \int_o^\infty [p(x,t) + q(x,t)] \, dt \qquad (103)$$

Equation 103 is easily evaluated experimentally when the solute concentration can be measured anywhere in the bed. The denominator is found by determining the concentration of solute in the void and penetrable volumes at position "x" (or some small interval centered on "x") at various evenly spaced times and summing the measurements. The numerator is found in a similar manner except that each concentration is multiplied by the time at which it is measured.

For a large zone experiment we find that

$$R_1(x) = (1 + K)(h - x)/u + K/[k_{-1}(1 + K)] + T/2u \qquad (104)$$

$$R_2(x) - R_1^2(x) = 2K(h - x)/(k_{-1}u) + K(2 + K)/[k_{-1}^2(1 + K)] + T^2/(12u^2) \qquad (105)$$

for the mean and variance of the residence time within $h - x$. Thus K and k_{-1} can be determined from the slopes and intercepts of the graphs of the mean and variance as a function of the distance $h - x$ down the bed. Equations 104 and 105 can be expressed in terms of a volume profile rather than a time profile in the usual way, by multiplying Equation 104 by the flow rate $F = V_o u/h$ and Equation 105 by F^2. If the measurements at position "x" are proportional to the total solute concentrations $p(x,t) + q(x,t)$ and are accurate, then this approach has the advantage that more data are obtained from each experiment. The kinetic characteristics of sorption-desorption must be known if column chromatography is to be extended to quantitative studies of chemical-reaction kinetics.

F. Relation to Plate Height Theory

Here we use the results of Sections C and D to study plate height (proportional to the dispersion in the profile divided by the square of the mean) as a function of the convection velocity "u". At very low "u", the plate height approaches a constant (rather than diverging as u^{-1}) which is explicitly given by the theory. As "u" increases, the plate height passes through a minimum in the usual way and then increases. We show that the past assumptions, that mass transfer and various types of diffusion contribute additively to the plate height, hold only at or beyond the minimum. We show further that our expression fits observed data with only a single adjustable-lumped parameter. The parameter contains the mass-transfer rate k_{-1} for moving out of the bead. The data fit as well as an analytical approximation that we derive for the position of the minimum, providing a relation between k_{-1} and bead particle size d_p and, consequently, between d_p and profile dispersion.

1. Definitions and Theory

Since plate height terminology is still widely used, we now present and discuss the results of previous sections in that notation. The spreading of solute molecules from a thin initial layer to a bell-shaped distribution along the bed at later times is called dispersion or zone spreading or band broadening. The causes of dispersion are sorption-desorption kinetics (also called nonequilibrium effects) and diffusion in the mobile phase. Here we use diffusion in the mobile phase as a general term which includes lateral (longitudinal) diffusion and diffusion-related flow phenomena such as eddy diffusion and velocity-profile heterogeneity.

The relative dispersion about the mean of the elution profile is the variance S_e divided by the square of the mean M_e. If diffusion above the bed is negligibly small, then from Equations 91 and 94 we obtain

$$\frac{S_e}{M_e^2} = \frac{\alpha r}{1 - (1 - e^{-r})/r} + \frac{2[1 + 2e^{-r} - 2(1 - e^{-r})/r - (1 - e^{-2r})/2r]}{r[(1 - e^{-r})/r]^2} \tag{106}$$

where $r = uh/D$ and

$$\alpha = 2KD/(h^2 k_{-1}(1 + K)^2) \tag{107}$$

The two terms in Equation 106 correspond to the two causes of dispersion described above, namely, nonequilibration of mass transfer and diffusion.

The height equivalent to a theoretical plate is defined as[3,33,46] $H = hS_e/M_e^2$. Division by the bead diameter d_p leads to the reduced plate height

$$\bar{h} = hS_e/(d_p M_e^2) \tag{108}$$

which is dimensionless. Using Equation 106 one can show that $\bar{h}$ equals 2 $(\alpha + 1/3)$ h/d_p when $r = 0$, that it decreases to a minimum when $r \cong \sqrt{2/\alpha}$ and that it then increases. Near and beyond the minimum, a good approximation is

$$\bar{h} = 2/v + \alpha h/d_p + (h/d_p)^2 v \tag{109a}$$

where $v = rd_p/h = ud_p/D$ is a dimensionless quantity called the reduced velocity. For typical parameter values and typical reduced velocities ($v \gtrsim 0.01$), the second term in Equation 109a is entirely negligible so that

$$\bar{h} \cong 2/v + \alpha(h/d_p)^2 v \tag{109b}$$

Equation 109a is similar to the van Deemter equation[3,46] given by

$$\bar{h} = B/v + A + Cv \tag{110}$$

where the constants B, A, and C are associated with axial molecular diffusion, eddy diffusion, and nonequilibrated mass transfer. However, Equation 110 involves three undetermined parameters, B, A, and C, while 109a has only one. Many equations similar to 110 have been derived for the reduced plate height.[42-44] For example, the Knox equation[44] is

$$\bar{h} = B/v + Av^{1/3} + Cv \tag{111}$$

Although Equation 106 is valid for $v \geq 0$, Equations 109, 110, and 111 are only reasonable near and beyond the minimum since the plate height should be a constant when $v = 0$ and should not be infinite.

2. Data Analysis

Plots of log $\bar{h}$ against log v have been obtained experimentally.[3,42,44] Using a data set from Figure 7 in Reference 42 we have determined the best fits of Equations 109b, 110, and 111. The best least-square fits of the two empirical Equations 110 and 111 are virtually identical in terms of root mean-square error, although the parameter sets (A, B, and C) are of course different for the two equations. Our Equation 109b, derived from a basic chromatography model, is not quite as accurate but it has only one adjustable parameter (Figure 2).

Interestingly, the best least-square parameter B in the Knox equation turns out to be 2 as predicted by our result. The strikingly close agreement is probably to some extent fortuitous. Nevertheless, the agreement suggests that the main reason for the difference between the data and the prediction based on Equation 109b is the absence of velocity profile effects. As a check on this we added a term $Av^{1/3}$ onto Equation 109b thus producing a two—parameter equation, and obtained a fit that was slightly better (in terms of RMS error) than that based on the Knox equation. One way of looking at these results is to regard Equation 109b as providing a partial theoretical basis for the Knox equation, pinning down two of the three adjustable parameters in that equation to values determined by a molecular model.

3. A Relation Between the Desorption Rate Constant and the Bead Diameter

Let v_m be the reduced velocity at which the minimum of $\bar{h}$ occurs. Since the minimum of Equation 106 occurs when $r \cong \sqrt{2/\alpha}$ where α is given by Equation 102 we obtain

$$k_{-1} = \frac{KDv_m^2}{(1 + K)^2 \, d_p^2} \tag{112}$$

From various plots of log $\bar{h}$ against log v, it has been observed[44] that v_m is consistently about 3. Thus, Equation 112 predicts that the desorption rate constant k_{-1} is inversely proportional to the square of the bead diameter d_p.

III. QUANTITATIVE AFFINITY CHROMATOGRAPHY

A. Competitive Elution When Molecules Bind Monovalently to Sites on Porous Beads
1. The Model

We now formulate an initial-boundary value problem that models a small-zone affinity

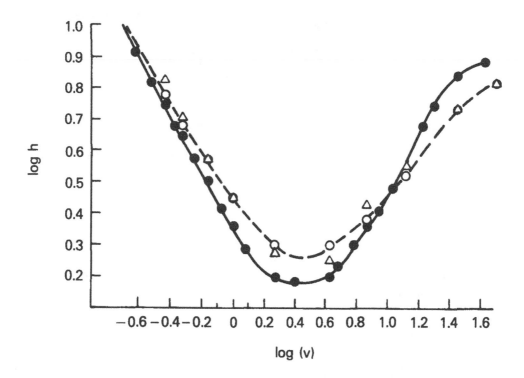

FIGURE 2.   Log-Log plot of reduced plate height h̄ as a function of reduced velocity ''v''. (-△-) are the experimental data points.[42] The RMS error of the fits using the van Deemter Equation 110 and the Knox Equation 111 are virtually identical (dashed line). The best least-square fit of Equation 109b is shown by the solid line and the curve was generated using the experimental value h/d_p = 3837 and determining $\alpha(h/d_p)^2$ = 0.26 by a least-squares fit.

chromatography experiment with a single type of molecule in the sample. We continue to consider a column with beads packed to a height ''h'' and molecules transported at velocity ''u'' in the mobile phase. In keeping with the results of our analysis in Section D, we assume that the flow rate is fast enough so that diffusion can be neglected. Assume that the molecule to be studied binds only monovalently to the ligand. The concentration of free competitive ligands in solution is L and the concentration of ligands that are covalently attached to the beads and available for binding molecules is N.

Let $p_1(x,t)$ and $p_2(x,t)$ be the probability-density functions (per unit column length) at position ''x'' at time ''t'' that molecules are in the mobile phase and are unbound and bound to free competitors, respectively. Let $q_1(x,t)$, $q_2(x,t)$, and $r(x,t)$ be the probability-density functions that molecules are inside the beads and unbound, bound to competitor, and bound to molecules covalently immobilized on the beads, respectively. We continue to use k_1 and k_{-1} as the rate constants for movement of molecules in and out of the beads (i.e., mass transfer). For association and dissociation with competitor we use rate constants k_2 and k_{-2}, and for association and dissociation with the bead-bound molecule, k_3 and k_{-3}. Equilibrium constants are defined by $K_i = k_i/k_{-i}$. The diagram below indicates the transfer rates between the five states

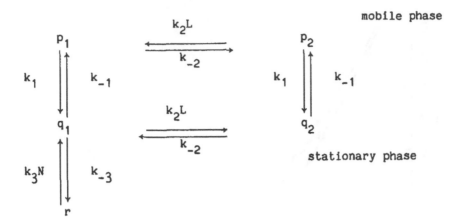

The size and mass of an eluted molecule is assumed to be large compared to those of the ligand so that its mass-transfer rates are unaffected when it binds a ligand.

From conservation of mass as described in Section II.B.1, we can derive the following system of reaction-transport partial-differential equations for the probability-density functions

$$\frac{\partial p_1}{\partial t} = u \cdot \frac{\partial p_1}{\partial x} - k_1 p_1 + k_{-1} q_1 - k_2 L p_1 + k_{-2} p_2 \tag{113}$$

$$\frac{\partial p_2}{\partial t} = u \cdot \frac{\partial p_2}{\partial x} - k_1 p_2 + k_{-1} q_2 + k_2 L p_1 - k_{-2} p_2 \tag{114}$$

$$\frac{\partial q_1}{\partial t} = k_1 p_1 - k_{-1} q_1 - k_2 L q_1 + k_{-2} q_2 - k_3 N q_1 + k_{-3} r \tag{115}$$

$$\frac{\partial q_2}{\partial q} = k_1 p_2 - k_{-1} q_2 + k_2 L q_1 - k_{-2} q_2 \tag{116}$$

$$\frac{\partial r}{\partial t} = k_3 N q_1 - k_{-3} r \tag{117}$$

These equations do not have diffusion-related terms since we have assumed that the flow rate, $F = uV_o/h$, is large enough so that the effects of diffusion can be neglected.

We confine attention to small-zone experiments and hence represent the initial layer by a Dirac-delta function. In previous formulations[9] of the monovalent-binding model, we have assumed that all of the molecules in the initial layer are unbound in the mobile phase ($p_1(x,0) = \delta(x - h)$, $p_2(x,0) = 0$). Here we make a different assumption, namely, that the molecules in the initial layer have equilibrated with the free ligand either before the initial layer is inserted or before the column flow is started. Thus the initial conditions are

$$p_1(x,0) = \frac{\delta(x - h)}{1 + K_2 L} \; ; \; p_2(x,0) = \frac{K_2 L \, \delta(x - h)}{1 + K_2 L}$$

$$q_1(x,0) = q_2(x,0) = r(x,0) = 0 \qquad x \neq h \tag{118}$$

The advantage of this new assumption is that the expressions for the mean and variance of the elution profile are simpler than those in Reference 9.

Since there are no molecules at the top of the bed after the initial layer leaves

$$p_1(h,t) = p_2(h,t) = q_1(h,t) = q_2(h,t) = r(h,t) = 0 \tag{119}$$

for $t > 0$. As in the previous sections we assume flow at the bottom of the bed, where the elution profile $u \cdot p(0,t)$ is measured, to be the same as if the bed extended below $x = 0$. Thus we solve the initial boundary value problem on the semi-infinite interval $(-\infty, h)$.

2. The Mean Elution Time From an Affinity Column

Although explicit solutions of the chromatography model in Equations 113 to 119 are not available, we can still obtain expressions for the moments by using a passage-time approach as in Sections II.C and D.

Recall that the mean-passage time, $T_1(x)$, is defined as the mean time for molecules starting in the initial layer to move past a position "x". The mean elution time, $M_e = T_1(0)$, is the mean time for molecules starting at $x = h$ to leave the bottom of the bed at $x = 0$. These definitions and those of higher moments can be expressed formally as

$$T_j(x) = \int_0^\infty t^j u[p_1(x,t) + p_2(x,t)] \, dt \tag{120}$$

Integration of Equations 113 to 117 with respect to "t" from 0 to ∞ yields the five equations below where all integrals are from 0 to ∞ and integrations are with respect to "t".

$$\frac{\partial}{\partial x} \int up_1 = k_1 \int p_1 - k_{-1} \int q_1 + k_2 L \int p_1 - k_{-2} \int p_2 - \frac{\delta(x-h)}{1 + K_2 L} \tag{121}$$

$$\frac{\partial}{\partial x} \int up_2 = k_1 \int p_2 - k_{-1} \int q_2 - k_2 L \int p_1 + k_{-2} \int p_2 - \frac{K_2 L \, \delta(x-h)}{1 + K_2 L} \tag{122}$$

$$0 = k_1 \int p_1 - k_{-1} \int q_1 - k_2 L \int q_1 + k_{-2} \int q_2 - k_3 N \int q_1 + k_{-3} \int r \tag{123}$$

$$0 = k_1 \int p_2 - k_{-1} \int q_2 + k_2 L \int q_1 - k_{-2} \int q_2 \tag{124}$$

$$0 = k_3 N \int q_1 - k_{-3} \int r \tag{125}$$

The solution of the five equations above using the conditions in Equation 119 are

$$\int p_1 = \frac{1}{u(1 + K_2 L)}, \int p_2 = \frac{K_2 L}{u(1 + K_2 L)} \tag{126}$$

$$\int q_1 = \frac{K_1}{u(1 + K_2 L)}, \int q_2 = \frac{K_1(K_2 L)}{u(1 + K_2 L)}, \int r = \frac{K_1 K_3 N}{u(1 + K_2 L)} \tag{127}$$

If the sum of Equations 113 to 117 is multiplied by "t" and integrated from 0 to ∞ with respect to "t", then after integration by parts, we obtain

$$\frac{\partial}{\partial x} \int tu(p_1 + p_2) = -\int (p_1 + p_2 + q_1 + q_2 + r) \tag{128}$$

Using the Definition Equation 120 and the Equations 126 and 127, we find

$$\frac{d}{dx} T_1(x) = -\left(1 + K_1 + \frac{K_1 K_3 N}{1 + K_2 L}\right) \frac{1}{u} \tag{129}$$

Integrating Equation 128 from 0 to "h", we obtain

$$M_e = T_1(0) = \left(1 + K_1 + \frac{K_1 K_3 N}{1 + K_2 L}\right) \frac{h}{u} \tag{130}$$

This expression for the mean-elution time M_e is simplier than the expression in Reference 9 because of the new assumption in Equation 118. Indeed, we can conclude that the exponential term in Equation 11 in Reference 9 is due to the nonequilibration of the molecules in the initial layer with the free ligand.

As in earlier sections we can convert the mean-elution time M_e to the mean-elution volume V_e by multiplying by the flow rate $F = V_o u/h$ and using $K_1 = V_p/V_o$. Thus the mean-elution volume for monovalent binding is

$$V_e = V_o + V_p\left(1 + \frac{K_3 N}{1 + K_2 L}\right) \tag{131}$$

Desorption rate constants k_{-1} seem to vary from approximately 0.01 sec^{-1} for larger beads to approximately 100 sec^{-1} for small beads. The reverse-rate constants such as k_{-3} for chemical reactions may range from 10^{-5} to 10^3 sec^{-1}.[47] Thus the chemical-reaction kinetics may be much faster or slower than the sorption-desorption (mass-transfer) kinetics. The Expressions 130 and 131 hold for both fast- and slow-chemical reactions. Because the general expression for the variance of the elution profile would be very complicated, we will find expressions for the variance in three special cases; fast-chemical reactions, slow-chemical reactions, and absence of competitive-free ligand.

3. Profile Variance for Fast-Chemical Reactions

If chemical-reaction kinetics are fast compared to the movement of molecules in and out of the beads, then a reasonable approximation is obtained if we assume that chemical equilibrium is attained in the mobile phase and in the stationary phase. In this case

$$p \equiv p_1 + p_2, \; p_1 = p/(1 + K_2 L), \; p_2 = pK_2 L/(1 + K_2 L) \tag{132}$$

$$s \equiv q_1 + q_2 + r, \; q_1 = \frac{s}{1 + K_2 L + K_3 N}, \; q_2 = \frac{sK_2 L}{1 + K_2 L + K_3 N},$$

$$r = \frac{sK_3 N}{1 + K_2 L + K_3 N} \tag{133}$$

so that the five partial-differential Equations 113 to 117 reduce to the two following equations

$$\frac{\partial p}{\partial p} = u \cdot \frac{\partial p}{\partial x} - k_1 p + \frac{k_{-1}(1 + K_2 L)}{1 + K_2 L + K_3 N} \cdot s \tag{134}$$

$$\frac{\partial s}{\partial t} = k_1 p - \frac{k_{-1}(1 + K_2 L)}{1 + K_2 L + K_3 N} \cdot s \tag{135}$$

Since this system of equations is analogous to that analyzed in Section II.D.3 we use the results there to find that the variance, S_e, is

$$S_e = \frac{2K_1}{k_{-1}} \left[1 + \frac{K_3N}{1 + K_2L} \right]^2 \frac{h}{u} \tag{136}$$

The variance W_e of the elution profile as a function of the volume eluted is found by multiplying S_e by F^2. The result is

$$W_e = F^2 S_e = \frac{2FV_p}{k_{-1}} \left(1 + \frac{K_3N}{1 + K_2L} \right)^2 \tag{137}$$

4. Profile Variance for Slow-Chemical Reactions

If chemical-reaction kinetics are slow compared to mass-transfer kinetics, then an approximation is obtained by assuming that mass-transfer equilibrium is attained for the unbound molecules and for the molecules bound to competitor. In this case

$$f \equiv p_1 + q_1 \quad p_1 = \frac{f}{1 + K_1} \quad q_1 = \frac{fK_1}{1 + K_1} \tag{138}$$

$$g \equiv p_2 + q_2 \quad p_2 = \frac{g}{1 + K_1} \quad q_2 = \frac{gK_1}{1 + K_1} \tag{139}$$

so that the five partial-differential Equations 113 to 117 reduce to the three following equations

$$\frac{\partial f}{\partial t} = \frac{u}{1 + K_1} \cdot \frac{\partial f}{\partial x} - k_2Lf + k_{-2}g - \frac{k_3NK_1}{1 + K_1} \cdot f + k_{-3}r \tag{140}$$

$$\frac{\partial g}{\partial t} = \frac{u}{1 + K_1} \cdot \frac{\partial g}{\partial x} + k_2Lf - k_{-2}g \tag{141}$$

$$\frac{\partial r}{\partial t} = \frac{k_3NK_1}{1 + K_1} \cdot f - k_{-3}r \tag{142}$$

The moments of the passage time are now

$$T_j(x) = \int_0^\infty t^j u(f + g)/(1 + K_1) \, dt \tag{143}$$

The second moment can be found by a procedure similar to that used in Section II.D.3. Integrating t^2 times the sum of Equations 140 to 142 yields

$$\frac{dT_2(x)}{dx} = -\frac{2(1 + K_1) T_1(x)}{u} - \frac{2K_1K_3N}{1 + K_1} \int tf - \frac{2K_1K_3N}{(1 + K_1) k_{-3}} \int f \tag{144}$$

The variance $S(x) = T_2(x) - T_1^2(x)$ satisfies

$$\frac{dS}{dx} = \frac{2K_1K_3N \, T_1(x)}{u(1 + K_2L)} - \frac{2K_1K_3N}{1 + K_1} \int tf - \frac{2K_1K_3N}{(1 + K_1) k_{-3}} \int f \tag{145}$$

If the differential equation for $\int_0^\infty tf\ dt$ is solved, then

$$\int tf = \frac{(1 + K_1)^2}{u^2(1 + K_2L)} \left[\left(1 + \frac{K_1K_3N}{(1 + K_1)(1 + K_2L)} \right) (h - x) + \frac{K_1K_3N\ u\ K_2L}{(1 + K_1)^2\ k_{-2}(1 + K_2L)^2} \right.$$

$$\left. (1 - e^{-B(h - x)/u}) \right]$$

where $B = (1 + K_1)(k_{-2} + k_2L)$. Thus Equation 145 becomes

$$\frac{dS}{dx} = \frac{-2K_1K_3N}{uk_{-3}(1 + K_2L)} - \frac{2(K_1K_3N)^2\ K_2L}{u(1 + K_1)\ k_{-2}(1 + K_2L)^3} (1 - e^{-B(h - x)/u}) \qquad (146)$$

Integrating Equation 146 from 0 to "h", we find

$$S_e = \frac{2K_1K_3N}{1 + K_2L} \left[\frac{1}{k_{-3}} + \frac{K_1K_2L\ K_3N}{(1 + K_1)\ k_{-2}(1 + K_2L)^2} \right] \frac{h}{u}$$

$$- \frac{2(K_1K_3N)^2\ K_2L}{(1 + K_1)^2\ k_{-2}^2(1 + K_2L)^4} (1 - e^{-Bh/u}) \qquad (147)$$

Although this expression for S_e is still not very simple, it is simpler than the expression obtained in Equation 34 in Reference 9. As before, the variance $W_e = F^2S_e$.

5. Profile Variance in the Absence of Competing-Free Ligand

When there is no free-competing ligand ($L = 0$) the five equations, 113 to 117, reduce to three equations for $p_1(x,t)$, $q_1(x,t)$, and $r(x,t)$. Following the procedure used in the previous section, we find

$$\frac{d}{dx} T_2(x) = -2 \int_0^\infty t(p_1 + q_1 + r)\ dt$$

$$= -2 \left[(1 + K_1 + K_1K_3N)^2 \frac{h - x}{u^2} + \frac{1}{k_{-1}} \frac{K_1(1 + K_3N)^2}{u} + \frac{K_1K_3N}{k_{-3}u} \right] \qquad (148)$$

and

$$\frac{d}{dx} S(x) = -2K_1 \left[\frac{(1 + K_3N)^2}{k_{-1}} + \frac{K_3N}{k_{-3}} \right] \frac{1}{u} \qquad (149)$$

Integrating Equation 149 from 0 to "h" leads to

$$S_e = 2K_1 \left[\frac{(1 + K_3N)^2}{k_{-1}} + \frac{K_3N}{k_{-3}} \right] \frac{h}{u} \qquad (150)$$

The variance W_e in terms of the volume eluted is

$$W_e = F^2S_e = 2FV_p \left[\frac{(1 + K_3N)^2}{k_{-1}} + \frac{K_3N}{k_{-3}} \right] \qquad (151)$$

6. The Effects of Heterogeneity

In many systems involving cellular receptors or membrane enzymes the molecules of interest are heterogeneous in their affinities and in their rate constants for ligand and a complete understanding of biological regulation requires quantitative characterization of this heterogeneity.[48-50] The effects of heterogeneity in the size, shape, and weight of the molecules and in the uniformity in packing, size, and structure of the beads were discussed in Section II.C.6. Here we assume that the physical properties of the molecules and the beads are uniform but the molecules are heterogeneous in their chemical reaction properties, such as their affinity for competitor. This heterogeneity is usually due to a distribution of reverse-rate constants. Since we have assumed that the concentration of molecules is small so that the chemical kinetics are linear, it follows that molecules with different equilibrium constants behave independently and that an average moment is the sum over the distribution of the separate moments.

For monovalent binding to molecules in porous beads, let $n(K_2, K_3)$ be the probability-density function for the equilibrium constants K_2 and K_3. Define the average of the "jth" moment of the passage time as

$$\overline{T}_j(x) = \int_o^\infty \int_o^\infty n(K_2, K_3) \int_o^\infty t \, up(x, t; K_2, K_3) \, dt dK_2 dK_3 \tag{152}$$

For fast-chemical reactions the average mean and variance are

$$\overline{M}_e = \left[1 + K_1 + K_1 \overline{\left(\frac{K_3 N}{1 + K_2 L} \right)} \right] \frac{h}{u}$$

$$\overline{S}_e = \frac{2K_1}{k_{-1}} \overline{\left(1 + \frac{K_3 N}{1 + K_2 L} \right)^2} \frac{h}{u} \tag{153}$$

For $K_2 L \gg 1$, we find

$$\frac{\overline{M}_e - (1 + K_1) h/u}{K_1 h/u} = \overline{\left(\frac{K_3}{K_2} \right)} \left(\frac{N}{L} \right) \tag{154}$$

Since K_2 and K_3 differ only in their diffusive parts, they are highly correlated so that $\overline{(K_3/K_2)}$ is approximately a constant.

For $K_2 L \ll 1$, we find

$$\frac{\overline{M}_e - (1 + K_1) h/u}{K_1 h/u} = \overline{K}_3 N \tag{155}$$

$$\overline{S}_e = \frac{2K_1 h}{k_{-1} u} \overline{(1 + K_3 N)^2} = \frac{2K_1 h}{k_{-1} u} \cdot \{ 1 + 2\overline{K}_3 N + N^2 [\overline{K}_3^2 + (\overline{K_3^2} - \overline{K}_3^2)] \} \tag{156}$$

Assuming K_1, k_{-1}, "h" and "u" are known, $\overline{K}_3$ could be obtained from measurements of M_e as a function of N by using Equation 155. Assuming K_1, k_{-1}, h, u, and $\overline{K}_3$ are known, the variance $\overline{K_3^2} - \overline{K}_3^2$ of the distribution of K_3 could be determined from measurements of $\overline{S}_e$ by using Equation 156. Thus, both the mean and variance of the distribution of K_3 can be obtained.

B. Competitive Elution When Molecules Bind Bivalently to Sites on Porous Beads

1. The Model

In this section a model is formulated for a small-zone affinity chromatography experiment with a homogeneous population of molecules with two identical binding sites. IgG- and IgA-myeloma proteins fall into this category. The molecules can first bind monovalently and then bivalently to sites on beads. As before, the concentration of free ligands in the solution is L and the concentration of ligands that are covalently attached to the beads and available for binding protein molecules is N.

Let $p_1(x,t)$, $p_2(x,t)$, and $p_3(x,t)$ be the probability-density functions at point "x" at time "t" that a molecule in the mobile phase has both sites empty, one site bound to a free ligand, or both sites bound to a free ligand, respectively. Let $q_1(x,t)$, $q_2(x,t)$, and $q_3(x,t)$ be the corresponding probability-density functions for molecules interior to a bead. Let $r_1(x,t)$, $r_2(x,t)$, and $r_3(x,t)$ be the probability-density functions that a molecule inside a bead has, respectively, one site empty and one bound to an attached molecule, one site bound to a free molecule and one bound to an attached molecule, and both sites bound to attached molecules. The transfer rates between the nine states are indicated in the diagram below

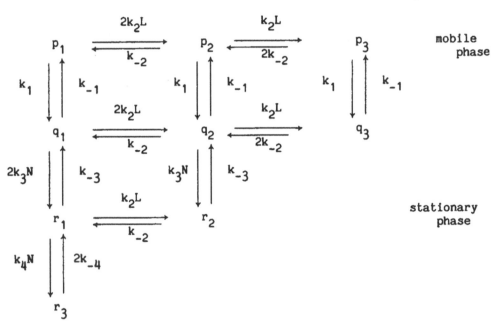

The mass-balance equations for this model are

$$\frac{\partial p_1}{\partial t} = u \cdot \frac{\partial p_1}{\partial x} - k_1 p_1 + k_{-1} q_1 - 2k_2 L p_1 + k_{-2} p_2 \tag{157}$$

$$\frac{\partial p_2}{\partial t} = u \cdot \frac{\partial p_2}{\partial x} - k_1 p_2 + k_{-1} q_2 + 2k_2 L p_1 - k_{-2} p_2 - k_2 L p_2 + 2k_{-2} p_3 \tag{158}$$

$$\frac{\partial p_3}{\partial t} = u \cdot \frac{\partial p_3}{\partial x} - k_1 p_3 + k_{-1} q_3 + k_2 L p_2 - 2k_{-2} p_3 \tag{159}$$

$$\frac{\partial q_1}{\partial t} = k_1 p_1 - k_{-1} q_1 - 2k_2 L q_1 + k_{-2} q_2 - 2k_3 N q_1 + k_{-3} r_1 \tag{160}$$

$$\frac{\partial q_2}{\partial t} = k_1 p_2 - k_{-1} q_2 + 2k_2 L q_1 - k_{-2} q_2 - k_2 L q_2 + 2k_{-2} q_3 - k_3 N q_2 + k_{-3} r_2 \quad (161)$$

$$\frac{\partial q_3}{\partial t} = k_1 p_3 - k_{-1} q_3 + k_2 L q_2 - 2k_{-2} q_3 \quad (162)$$

$$\frac{\partial r_1}{\partial t} = 2k_3 N q_1 - k_{-3} r_1 - k_2 L r_1 + k_{-2} r_2 - k_4 N r_1 + 2k_{-4} r_3 \quad (163)$$

$$\frac{\partial r_2}{\partial T} = k_3 N q_2 - k_{-3} r_2 + k_2 L r_1 - k_{-2} r_2 \quad (164)$$

$$\frac{\partial r_3}{\partial t} = k_4 N r_1 - 2k_{-4} r_3 \quad (165)$$

Here we also assume that the molecules in initial layers have equilibrated with the free ligand before the column flow is started. Thus the initial conditions are

$$p_1(x,0) = \frac{\delta(x - h)}{(1 + K_2 L)^2}, \ p_2(x,0) = \frac{2K_2 L \ \delta(x - h)}{(1 + K_2 L)^2} \ p_3(x,0) = \frac{(K_2 L)^2 \ \delta(x - h)}{(1 + K_2 L)^2}$$

$$q_1(x,0) = q_2(x,0) = q_3(x,0) = 0$$

$$r_1(x,0) = r_2(x,0) = r_3(x,0) = 0 \quad (166)$$

These initial conditions are different from those used in Reference 9 and they lead to simpler expressions for the moments. The boundary conditions corresponding to having no molecules at the top of the bed for time t > 0 are

$$p_1(h,t) = p_2(h,t) = p_3(h,t) = 0$$

$$q_1(h,t) = q_2(h,t) = q_3(h,t) = 0$$

$$r_1(h,t) = r_2(h,t) = r_3(h,t) = 0 \quad (167)$$

for t > 0. The "jth" moment of the elution profile is

$$T_j(x) = \int_0^\infty t^j u[p_1(x,t) + p_2(x,t) + p_3(x,t)] \ dt \quad (168)$$

2. The Mean Elution Time From an Affinity Column

In order to find the mean-passage time $T_1(x)$, and the mean-elution time, $M_e = T_1(0)$, we follow the same procedure as used for monovalent binding in Section II.A.2. Integration of the Equations 157 to 165 with respect to "t" from 0 to ∞ leads to nine equations with the first three being differential equations. By using the boundary conditions from Equation 167 we find the following solution

$$\int p_1 = \frac{1}{u(1 + K_2 L)^2}, \int p_2 = \frac{2K_2 L}{u(1 + K_2 L)^2}, \int p_3 = \frac{(K_2 L)^2}{u(1 + K_2)^2} \quad (169)$$

$$\int q_1 = \frac{K_1}{u(1 + K_2L)^2}, \int q_2 = \frac{2K_1K_2L}{u(1 + K_1L)^2}, \int q_3 = \frac{K_1(K_2L)^2}{u(1 + K_2L)^2} \qquad (170)$$

$$\int p_1 = \frac{2K_1K_3N}{u(1 + K_2L)^2}, \int p_2 = \frac{2K_1K_2L\ K_3N}{u(1 + K_2L)^2}, \int_3 p = \frac{K_1K_3NK_4N}{u(1 + K_2L)^2} \qquad (171)$$

Note that all integrals above are from 0 to ∞ with respect to "t". If the sum of Equations 157 to 165 are multiplied by "t" and integrated from 0 to ∞ with respect to "t", then after integration by parts and using the Equations 169 to 171 we find

$$\frac{dT_1(x)}{dx} = -\int (p_1 + p_2 + p_3 + q_1 + q_2 + r_1 + r_2 + r_3)$$

$$= -\left[1 + K_1 + \frac{2K_1K_3N}{1 + K_2L} + \frac{K_1K_3N\ K_4N}{(1 + K_2L)^2}\right]\frac{1}{u} \qquad (172)$$

Integrating Equation 172 from 0 to "h" we obtain

$$M_e = T_1(0) = \left(1 + K_1 + \frac{2K_1K_3N}{1 + K_2L} + \frac{K_1K_3N\ K_4N}{(1 + K_2L)^2}\right)\frac{h}{u} \qquad (173)$$

As usual, we can convert Equation 173 to an equation for the mean-elution volume by multiplying by the flow rate $F = V_ou/h$ and using $K_1 = V_p/V_o$. Thus the mean-elution volume for bivalent binding is

$$V_e = V_o + V_p\left[1 + \frac{2K_3N}{1 + K_2L} + \frac{K_3N\ K_4N}{(1 + K_2L)^2}\right] \qquad (174)$$

Because the general expression for the variance of the elution profile would be extremely complicated we will find expressions for the variance in two special cases, fast-chemical reactions and absence of competitive-free ligand.

3. Profile Variance for Fast Chemical Reactions

If the chemical reactions are fast compared to the movement of molecules in and out of beads, then a reasonable approximation is obtained if chemical equilibrium in the mobile phase and in the stationary phase is assumed. In this case

$$p \equiv p_1 + p_2 + p_3, p_1 = \frac{p}{(1 + K_2L)^2}, p_2 = \frac{p2K_2L}{(1 + K_2L)^2},$$

$$p_3 = \frac{p(K_2L)^2}{(1 + K_2L)^2} \qquad (175)$$

$$s \equiv q_1 + q_2 + q_3 + r_1 + r_2 + r_3,$$

$$q_1 = \frac{s}{(1 + K_2L)^2 + 2K_3N(1 + K_2L) + K_3NK_4N} \qquad (176)$$

$$q_2 = 2K_2Lq_1, q_3 = (K_2L)^2\ q_1, r_1 = 2K_3Nq_1, r_2 = K_2Lr_1, r_3 = K_4Nr_1/2 \qquad (177)$$

so that the nine partial differential equations reduce to the following two equations

$$\frac{\partial p}{\partial t} = u \cdot \frac{\partial p}{\partial x} - k_1 p + k_{-1} \cdot \frac{(1 + K_2 L)^2}{(1 + K_2 L)^2 + 2K_3 N(1 + K_2 L) + K_3 N K_4 N} \cdot s \quad (178)$$

$$\frac{\partial s}{\partial t} = k_1 p - k_{-1} \cdot \frac{(1 + K_2 L)^2}{(1 + K_2 L)^2 + 2K_3 N(1 + K_2 L) + K_3 N K_4 N} \cdot s \quad (179)$$

Since these differential equations are similar to those analyzed in Section II.D.3, we can use the results there to obtain

$$S_e = \frac{2K_1}{k_{-1}} \frac{h}{u} \left[1 + \frac{2K_3 N}{1 + K_2 L} + \frac{K_3 N K_4 N}{(1 + K_2 L)^2} \right]^2 \quad (180)$$

The corresponding formula for the variance of the elution profile when expressed as a function of the volume eluted is

$$W_e = F^2 S_e = \frac{2FV_p}{k_{-1}} \left[1 + \frac{2K_3 N}{1 + K_2 L} + \frac{K_3 N K_4 N}{(1 + K_2 L)^2} \right]^2 \quad (181)$$

4. Profile Variance in the Absence of Competing-Free Ligand

When there is no free-competing ligand ($L = 0$), the nine equations, 157 to 165, reduce to four equations for $p_1(x,t)$, $q_1(x,t)$, $r_1(x,t)$, and $r_3(x,t)$. Following the procedure used in the previous section, we obtain

$$\frac{dT_2(x)}{dx} = -2 \int_0^\infty t(p_1 + q_1 + r_1 + r_3) \, dt$$

$$= -2 \frac{T_1(x)}{u} \left[1 + K_1 + 2K_1 K_3 N + K_1 K_3 N K_4 N \right]$$

$$-2 \frac{K_1}{uk_{-1}} \left[1 + 2K_3 N + K_3 N K_4 N \right]^2$$

$$-2 \frac{2K_1 K_3 N}{uK_{-3}} \left[1 + \frac{K_4 N}{2} \right]^2 - 2 \frac{K_1 K_3 N K_4 N}{2uk_{-4}} \quad (182)$$

and since $S(x) = T_2(x) - T_1^2(x)$, we find

$$\frac{dS}{dx} = -\frac{2K_1}{u} \left[\frac{(1 + 2K_3 N + K_3 N K_4 N)^2}{k_{-1}} + \frac{2K_3 N(1 + K_4 N/2)^2}{k_{-3}} + \frac{K_3 N K_4 N}{2k_{-4}} \right] \quad (183)$$

Integrating Equation 183 from 0 to "h" leads to

$$S_e = \frac{2K_1 h}{u} \left[\frac{(1 + 2K_3 N + K_3 N K_4 N)^2}{k_{-1}} + \frac{2K_3 N(1 + K_4 N/2)^2}{k_{-3}} + \frac{K_3 N K_4 N}{2k_{-4}} \right] \quad (184)$$

The variance in terms of the volume eluted is

$$W_e = 2FV_p \left[\frac{(1 + 2K_3 N + K_3 N K_4 N)^2}{k_{-1}} + \frac{2K_3 N(1 + K_4 N/2)^2}{k_{-3}} + \frac{K_3 N K_4 N}{2k_{-4}} \right] \quad (185)$$

C. Competitive Elution When Molecules Bind Monovalently to Ligands on Impenetrable Beads

1. The Model

The model here is similar to previous models described except that the beads are impenetrable and the molecules are covalently attached to their surfaces. Consider a monovalent molecule and let $p_1(x,t)$, $p_2(x,t)$, and $r(x,t)$ be the probability-density functions that it is unbound, (in the mobile phase), bound to a free ligand (in the mobile phase), and has its site bound to a bead molecule, respectively. The transfer diagram is

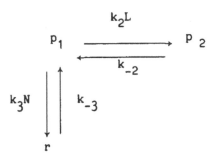

The mass-balance equations are

$$\frac{\partial p_1}{\partial t} = u \cdot \frac{\partial p_1}{\partial x} - k_2 L p_1 + k_{-2} p_2 - k_3 N p_1 + k_{-3} r \tag{186}$$

$$\frac{\partial p_2}{\partial t} = u \cdot \frac{\partial p_2}{\partial X} + k_2 L p_1 - k_{-2} p_2 \tag{187}$$

$$\frac{\partial r}{\partial t} = k_3 N p_1 - k_{-3} r \tag{188}$$

As in Sections III.A.1 and III.B.1 we assume that the molecules in the initial layer have equilibrated with the free ligand before the column flow is started. This assumption is not the same as in Reference 9 so we have different initial conditions and simpler resulting formulas. The initial conditions corresponding to the thin-initial layer are

$$p_1(x,0) = \frac{\delta(x - h)}{1 + K_2 L}, \; p_2(x,0) = \frac{K_2 L \, \delta(x - h)}{1 + K_2 L} , \; r(x,0) = 0 \qquad x \neq h \tag{189}$$

Since there are no molecules at the top of the bed after the initial layer leaves, we have for $t > 0$

$$p_1(h,t) = p_2(h,t) = r(h,t) = 0 \tag{190}$$

Here the elution profile is given by $u[(p_1(0,t) + p_2(0,t)]$.

2. The Mean and Variance of the Elution Profile

The Equations 186 to 188 are similar to Equations 140 to 142 so that we can convert the results there to obtain

$$M_e = \left[\left(1 + \frac{K_3 N}{1 + K_2 L} \right) \frac{h}{u} \right] \tag{191}$$

$$S_e = \frac{2K_3N}{1 + K_2L} \left[\frac{1}{k_{-3}} + \frac{K_2LK_3N}{k_{-2}(1 + K_2L)^2} \right] \frac{h}{u}$$

$$- \frac{2(K_2L)(K_3N)^2}{k_{-2}^2(1 + K_2L)^4} (1 - e^{-Bh/u}) \tag{192}$$

where $B = (k_{-2} + k_2L)$.

When there is no free competing molecule $(L = 0)$, these expressions simplify to

$$M_e = (1 + K_3N) h/u \tag{193}$$

$$S_e = \frac{2K_3N}{k_{-3}} \cdot \frac{h}{u} \tag{194}$$

As before, the results of Equations 191 and 192 can be converted to the mean-elution volume, variance of the elution volume, and variance of the elution profile as a function of volume eluted by using the flow rate $F = V_o u/h$. We obtain

$$V_e = V_o \left(1 + \frac{K_3N}{1 + K_2L} \right) \tag{195}$$

$$W_e = \frac{2FV_oK_3N}{1 + K_2L} \left[\frac{1}{k_{-3}} + \frac{K_2LK_3N}{k_{-2}(1 + K L)^2} \right] - \frac{2F^2K_2L(K_3N)^2}{k_{-2}^2(1 + K_2L)^4} (1 - e^{-BV_o/F}) \tag{196}$$

3. Comparison With a Previous Theory

Denizot and Delaage[32] used the random-walk approach of Giddings and Eyring[34] to develop a theory of affinity chromatography for monovalent binding of molecules to beads. Their expression for the mean of the elution profile is

$$\text{mean} = E(T') = E(t_o)(1 + k/k') \tag{197}$$

where $E(t_o)$ is the expected elution time for a particle that does not bind, and k/k' is the equilibrium constant of binding (K_3N in our notation). For porous beads $E(t_o) = (1 + K_1)h/u$ so that Equation 197 in our notation becomes

$$M_e = (1 + K_1)(1 + K_3N) h/u \tag{198}$$

This is clearly not the same as Equation 130 with $L = 0$ so that their theory does not seem to be applicable to porous beads. This is consistent with Chaiken's finding[5] of unrealistic rate constants when the Denizot and Delaage theory was applied to the elution of ribonuclease through a bed of porous beads. The difficulty, as they indicate, is that their formulas assume that mass transfer, in and out of the stationary phase, is insignificant. In fact, for impenetrable beads $E(t_o) = h/u$ so that Equation 197 is the same as our result, Equation 193. The expression of Denizot and Delaage for the variance is

$$\sigma'^2 = \frac{2k}{k'^2} \cdot E(t_o) + \left(1 + \frac{k}{k'} \right)^2 \sigma_o^2 \tag{199}$$

where σ_o^2 is the variance of t_o. For impenetrable beads $E(t_o) = h/u$ and $\sigma_o^2 = 0$ when the effects of diffusion are negligible so that Equation 199 in our notation becomes Equation 194. Thus their results do apply to chromatography with impenetrable beads.

It is important to recognize, however, that the forward- and reverse-rate constants obtained when one of the reactants is attached to a large particle, such as a bead or a cell, may differ by several orders of magnitude from the values obtained when both reactants are dispersed in solution.[14,23] The differences are present[32] when the reactions are diffusion limited and are the consequences of the Brownian movement of molecules in the presence of a large surface (e.g., a bead) as discussed in Section A. If this is the case, rate constants obtained by chromatography experiments would have to be interpreted entirely differently from rate constants obtained in well-stirred solutions and quantitative agreement between the two would be fortuitous and surprising. Nevertheless, rate constants obtained by chromatography, in which flow is a dominant process, may be more relevant to physiological situations, where flow is also dominant, than well-stirred solution data.

D. Competitive Elution When Molecules Bind Bivalently to Ligands on Impenetrable Beads

1. The Model

The model here is similar to the model in Section III.B.1 except that the beads are impenetrable. Using the same notation, the transfer diagram is

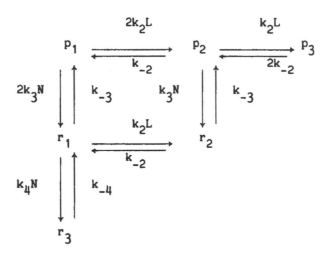

The mass-balance equations for this model are

$$\frac{\partial p_1}{\partial t} = u \cdot \frac{\partial p_1}{\partial x} - 2k_2 L p_1 + k_{-2} p_2 - 2k_3 N p_1 + k_{-3} r_1 \tag{200}$$

$$\frac{\partial p_2}{\partial t} = u \cdot \frac{\partial p_2}{\partial x} + 2k_2 L p_1 - k_{-2} p_2 - k_2 L p_2 + 2k_{-2} p_3 - k_3 N P_2 + k_{-3} r_2 \tag{201}$$

$$\frac{\partial p_3}{\partial t} = u \cdot \frac{\partial p_3}{\partial x} + k_2 L p_2 - 2k_{-2} p_3 \tag{202}$$

$$\frac{\partial r_1}{\partial t} = 2k_3 N p_1 - k_{-3} r_1 - k_2 L r_1 + k_{-2} r_2 - k_4 N r_1 + 2k_{-4} r_3 \tag{203}$$

$$\frac{\partial r_2}{\partial t} = k_2 L r_1 - k_{-2} r_2 + k_3 N p_2 - k_{-3} r_2 \tag{204}$$

$$\frac{\partial r_3}{\partial t} = k_4 N r_1 - 2k_{-4} r_3 \tag{205}$$

The initial and boundary conditions are analogous to those in the previous section.

2. The Moments of the Elution Profile

Using the procedure developed in previous sections, we find that the mean-elution time is

$$M_e = \frac{h}{u} \left[1 + \frac{2K_3 N}{1 + K_2 L} + \frac{K_3 N K_4 N}{(1 + K_2 L)^2} \right] \tag{206}$$

The mean elution volume is

$$V_e = V_o \left[1 + \frac{2K_3 N}{1 + K_2 L} + \frac{K_3 N K_4 N}{(1 + K_2 L)^2} \right] \tag{207}$$

When there is no competing-free ligand (L = 0) the six equations, 200 to 205, reduce to three differential equations. Since this situation is formally analogous to monovalent binding with porous beads and L = 0, we can convert the variance formulas in Section III.A.5 to obtain

$$S_e = \frac{4K_3 Nh}{uk_{-3}} \left(1 + \frac{K_4 N}{2} \right)^2 + \frac{K_3 N K_4 Nh}{uk_{-4}} \tag{208}$$

and

$$W_e = \frac{4FV_o K_3 N}{k_{-3}} \left(\frac{1 + K_4 N}{2} \right)^2 + \frac{FV_o K_3 N K_4 N}{k_{-4}} \tag{209}$$

E. Competitive Elution When Substrate Binds to Macromolecules on Porous Beads

In contrast to previous sections in which macromolecules were eluted, and the bead-bound ligands were relatively small, we now consider competitive affinity chromatography when the roles of the macromolecule and the ligand are reversed, i.e., the macromolecules are now covalently attached to the porous beads and the substrate (ligand) is eluted.[51,52] In this reversed-role affinity chromatography, competition is different since the soluble inhibitor now competes with the substrate for binding sites on the macromolecules attached to the beads.

1. The Model

Assume that the macromolecules are covalently attached to the beads, that a small amount of substrate starts in a thin layer at the top of the bed, and that soluble inhibitor is uniformly distributed throughout the column. Since the small substrate molecules have a larger diffusion constant and move in and out of the beads more easily than the macromolecules, the desorption rate constant k_{-1} will be larger.[3] In this case there will be a smaller contribution to the profile variance from sorption-desorption kinetics so that estimation of the chemical-dissociation rate constant might be possible.

Consider a cylindrical column packed to a height "h" with permeable beads. Let N be the concentration of binding sites on bead-attached macromolecules that are available for binding soluble substrate or soluble inhibitor; this should be nearly all of the sites on the macromolecules since the small substrate and inhibitor molecules can penetrate the entire bead. The concentration N could be estimated by measuring the amount of macromolecules used in constructing the beads or by counting substrate molecules washed off after saturation of the binding sites.

Before the zonal elution experiment, the chemical reaction between the inhibitor with concentration I and the binding sites with concentration N will be at equilibrium since the inhibitor molecules are uniformly distributed throughout the column. Thus the concentration of sites on macromolecules filled with inhibitor will be K_2IN where K_2 is the equilibrium constant. Assume that the concentration S of substrate molecules is always sufficiently smaller than the inhibitor concentration I and the binding site concentration N so that the effect of the substrate on the macromolecule inhibitor equilibrium is negligible, and the concentration N* of free sites available for binding substrate satisfies $N* = N/(1 + K_2I)$. Since the concentration S is small, reaction kinetics with the free sites on the bead-bound macromolecules will be linear.

Following the approach in previous sections, we let p(x,t), q(x,t), and r(x,t) be the probability densities at distance "x" from the bottom of the bed at time "t" of substrate in the mobile phase, of substrate unbound in the stationary phase, and of substrate bound in the stationary phase, respectively. The transfer diagram is

$$
p \; \underset{k_{-1}}{\overset{k_1}{\rightleftharpoons}} \; q \; \underset{k_{-3}}{\overset{k_3N^*}{\rightleftharpoons}} \; r
$$

where k_1, k_{-1}, and $K_1 = k_1/k_{-1}$ are the sorption rate constant, desorption rate constant, and mass transfer equilibrium constant, respectively; and k_3, k_{-3}, and $K_3 = k_3/k_{-3}$ are the chemical association rate constant, dissociation rate constant, and equilibrium constant, respectively. The flow rate is assumed to be large enough so that the effects of diffusion are negligible.

The small-zone chromatography experiment can be described by the mass-balance equations and the initial and boundary conditions given below.

$$\frac{\partial p}{\partial t} = u \frac{\partial p}{\partial x} - k_1 p + k_{-1}q \tag{210}$$

$$\frac{\partial q}{\partial t} = k_1 p - k_{-1}q - k_3 N^* q + k_{-3} r \tag{211}$$

$$\frac{\partial r}{\partial t} = k_3 N^* q - k_{-3}r \tag{212}$$

$$p(x,0) = \delta(x - h) \tag{213}$$

$$q(x,0) = r(x,0) = 0 \qquad x \neq h \tag{214}$$

$$p(h,t) = q(h,t) = r(h,t) = 0 \qquad t > 0 \tag{215}$$

As usual, "u" is the mobile phase velocity, δ is the Dirac delta function, $F = uV_o/h$ is the flow rate, and up (0,t) is the elution profile.

2. The Moments of the Elution Profile

Although explicit solutions of the chromatography model given by Equations 210 to 215 are not available, the mean and variance of the passage time can be found by following the procedure used in previous sections. First the partial-differential equations, 210 to 212, are integrated from 0 to ∞ with respect to "t" and then the ordinary differential equation for p(x,t) is integrated from "x" to "h". Then "t" times the Equations 210 to 212 are integrated form 0 to ∞ with respect to "t". Algebraic manipulation of the six equations obtained leads to

$$\frac{dT_1(x)}{dx} = -[1 + K_1 + K_1K_3N^*]/u \qquad T_1(h) = 0 \qquad (216)$$

The first moment $T_1(x)$ can be found by integrating Equation 216 from "x" to "h" so that the mean-elution time is

$$M_e = T_1(0) = (1 + K_1 + K_1K_3N^*)(h/u). \qquad (217)$$

Next, t^2 times the Equations 210 to 212 are integrated from 0 to ∞ with respect to "t". Algebraic manipulation of these equations combined with previous equations leads to the following differential equation for the variance $S(x) = T_2(x) - T_1^2(x)$

$$\frac{dS(x)}{dx} = -2(1 + K_3N^*)^2 K_1/k_{-1}u - 2K_1K_3N^*/k_{-3}u, \quad S(h) = 0 \qquad (218)$$

Integrating Equation 218 from "x" to "h" yields $S(x)$ so that the variance of the elution profile is

$$S_e = S(0) = \frac{2K_1h}{u}\left[\frac{(1 + K_3N^*)^2}{k_{-1}} + \frac{K_3N^*}{k_{-3}}\right] \qquad (219)$$

Chemical-reaction kinetics are said to be slow compared to the mass-transfer kinetics when

$$k_{-3}(1 + K_3N^*)^2/(K_3N^*) << k_{-1} \qquad (220)$$

For slow-chemical kinetics the first term in Equation 219 is negligible so that the variance of the elution profile is predominantly due to local nonequilibration of chemical-reaction kinetics.

If the elution profile is given as a function of the volume eluted instead of time, then with void volume V_o, penetrable volume V_p, flow rate $F = V_ou/h$, and equilibrium constant $K_1 = V_p/V_o$, the mean-elution volume is

$$V_e = FM_e = V_o + V_p + V_pK_3N^* \qquad (221)$$

and the variance of the elution profile is

$$W_e = F^2S_e = 2FV_p\left[\frac{(1 + K_3N^*)^2}{k_{-1}} + \frac{K_3N^*}{k_{-3}}\right] \qquad (222)$$

where $N^* = N/(1 + K_2I)$ is the concentration of free-binding sites.

F. Estimating Chemical-Reaction Equilibrium Constants From Elution Profiles

As macromolecules move through a chromatography column they interact with soluble ligand and also with ligand immobilized on beads. The elution profile depends on mass-transfer properties, such as flow and movement in and out of beads; on the free and bead-bound ligand concentrations; and on the chemical-reaction properties of the macromolecules with the free and bead-bound ligand. Thus it should be possible to estimate chemical-reaction constants from elution profiles. Dunn and Chaiken[53,54] developed a method for estimating equilibrium (dissociation) constants for monovalent proteins by affinity chromatography assuming that sorption-desorption kinetics and chemical-reaction kinetics are both at equilibrium locally. This method was extended to bivalent macromolecules by Chaiken, Eilat, and McCormick.[55,56] See Chaiken[5] for a review of these quantitative theories and their applications.

In this section we present and explain various procedures for estimating chemical-reaction equilibrium constants from means of elution profiles.

1. Macromolecules Binding to Ligands on Porous Beads

In Sections II.C and D a theory of small-zone competitive elution chromatography is presented which does not assume local equilibration of sorption-desorption kinetics. The variance of the elution profile in that model is related to the height equivalent to a theoretical plate, and criteria on the flow rate are given which when met, a guarantee that the effects of diffusion on the moments of the profile can be neglected. A procedure is given in Section II.D.5 for determining the sorption-desorption equilibrium constant K_1 from the elution profile. From the void volume V_o and the mean-elution volume V_e, the penetrable volume is $V_p = V_e - V_o$ and the mass-transfer equilibrium constant is $K_1 = V_p/V_o$. Approximate treatments of bead heterogeneity, molecular heterogeneity, and nonlinear kinetics, are developed in Sections II.C.5 and 6.

In Section III.A this theory is extended to affinity chromatography without assuming local equilibration of the interaction of macromolecules with bead-bound or free ligands. The parameter K_2 is the equilibrium (association) constant characterizing the interaction between macromolecules and free ligands, K_3 is the equilibrium constant with bead-bound ligands, L is the concentration of free ligands, and N is the concentration of available bead-bound ligands. The mean-elution volume V_e satisfies Equation 131 so that

$$\frac{V_p}{V_e - (V_o + V_p)} = \frac{1}{K_3 N} + \frac{K_2}{K_3 N} L \tag{223}$$

Thus $K_3 N$ and K_2 can be determined from the intercept and slope obtained by fitting Equation 223 to data using linear regression of the left side on L.

For macromolecules that bind bivalently, let K_4 be the equilibrium (association) constant for bivalent binding given that the macromolecule is already singly bound. Then, using Equation 174, we find that the mean-elution volume V_e satisfies

$$\frac{V_e - (V_o + V_p)}{V_p} = \frac{2K_3 N}{1 + K_2 L} + \frac{K_3 N \, K_4 N}{(1 + K_2 L)^2} \tag{224}$$

Nonlinear regression can be applied to Equation 224 using data pairs consisting of L and the corresponding V_e to estimate K_2, $K_3 N$, and $K_4 N$.

Alternatively, Equation 224 can be manipulated to give

$$\frac{(V_e - V_o - V_p)(1 + K_2 L)}{V_p} = 2K_3 N + K_3 N \, K_4 N \frac{1}{1 + K_2 L} \tag{225}$$

Then the value of K_2 can be found which gives the best fit of Equation 225 to the data by linear regression of the left side on $(1 + K_2L)^{-1}$. Also the ratio of K_4 to K_3 can be found from the slope divided by the square of the intercept. This approach has been used by Inman[57] on mouse IgA myeloma protein to determine a value of K_2 that is within a few percent of the value obtained by equilibrium dialysis, and to obtain the ratio K_4/K_3. If the concentration N of available sites on bead-bound macromolecules were known then K_3 and K_4 could also be estimated rather than just their ratio.

One method for estimating the concentration N of available binding sites is to saturate the sites with labeled molecules by putting an excess of labeled molecules in the column. Next the nonbound molecules are washed out. Then the bound molecules are displaced by unlabeled molecules in the solution and washed out. The total number of bound molecules washed out is equal to the number of available binding sites if the molecules were bound monovalently. The number N of available binding sites is usually less than 5% of the total number of ligands in beads since most ligands are not accessible to the molecules.

2. Peaks and Means of the Elution Profile

Our theory of elution chromatography leads, when diffusion is negligible, to a mean-elution time $M_e = (1 + K_1)h/u$ where the equilibrium constant is $K_1 = V_p/V_o$. Since the volume eluted up to time "t" is the flow rate, $F = V_o u/h$, multiplied by the time "t", our formula for the profile mean in terms of the volume eluted is $V_e = V_o + V_p$. This formula is formally the same as a standard formula for the peak of the elution profile. The conditions under which this formula is applicable and comparisons of the peaks and the means are given in Section II.B.5.b.

Using the theory for monovalent binding to ligands immobilized in porous beads, which assumes local equilibration of both mass-transfer and chemical kinetics, Dunn and Chaiken[54] obtained the following important relation

$$\frac{1}{V - V_o} = \frac{1 + L/K_L}{(V_o - V_m)[LM]/K \, [LM]} \tag{226}$$

where V is the peak of the elution profile, V_o is the volume at which the protein elutes in the absence of interaction ($V_o + V_p$ in our notation), V_m is the void volume (V_o in our notation), [LM] is the concentration of immobilized molecule (N in our notation), K_{LM} is the dissociation constant for interaction of molecules with immobilized molecule ($1/K_3$ in our notation), [L] is the concentration of free molecule (L in our notation), and K is the dissociation constant for interaction of molecules with free molecule ($1/K_2$ in our notation). Thus in our notation their formula becomes

$$V = V_o + V_p + V_p \cdot \frac{K_3 N}{1 + K_2 L} \tag{227}$$

Hence their Formula 227 for the peak is formally the same our Formula 131 for the mean of the elution profile.

Since explicit solutions of the model in Equations 113 to 119 are not available, it is not possible to assess the difference between the peak and the mean as we did for elution chromatography in Section II.B.5.b. Of course, the peak and the mean are equal when the elution profile is symmetric and their difference increases as the elution profile becomes less symmetric. We conclude that the Equation 227 for the peak is only a reasonable approximation when diffusion is negligible, when the molecules in the initial layer are nearly equilibrated with the free ligand before flow started, and when the profile is nearly symmetric.

Chaiken et al.[56] have used a similar local-equilibrium model to derive a formula for bivalent binding to molecule attached to porous beads. In our notation their formula for the peak V becomes

$$V = V_o + V_p + V_p \left[\frac{2K_3N}{1 + K_2L} + \frac{(K_3N)^2}{(1 + K_2L)^2} \right] \qquad (228)$$

Their Formula 228 for the peak of the elution profile is almost the same as our Formula 174 for the mean of the elution profile. The difference is that we distinguish between the equilibrium constant K_3 for the binding of the molecule to a first ligand and the equilibrium constant K_4 for the binding of the molecule to a second ligand while they do not distinguish between these two equilibrium constants. Inman[57] found that for mouse IgA myeloma protein K_4 is approximately 5 times K_3.

3. Flow Rate Considerations

In order to determine equilibrium constants using affinity chromatography, the flow rate F in the column must satisfy certain criteria. First, the flow rate must be fast enough so that the effects of axial diffusion in the column can be neglected. Conditions which guarantee that axial diffusion is negligible are given in Section II.D.4. Diffusion can usually be neglected unless the flow rates are extremely slow. Second, the flow rate must be slow enough so that the molecules have many opportunities to move inside the beads and to react chemically with the ligand attached to the beads. If the flow rate is too fast, the molecules will rarely move inside a bead and may only interact chemically with those ligands that are on the surfaces of the beads.

It is possible to calculate the average number of times that a typical molecule moves into and out of a bead as it traverses the chromatography column. It is also possible to calculate the average number of times that a typical molecule binds to and detaches from ligands covalently attached to beads. We now indicate how to do these calculations for monovalent binding of molecules to ligands on porous beads.

Since $F = V_o u/h$ and $K_1 = V_p/V_o$, the Formula 130 for the mean-elution time can be written as

$$M_e = \frac{V_o}{F} + \frac{V_o K_1}{F} + \frac{V_p K_3 N}{F(1 + K_2 L)} \qquad (229)$$

The first term in Equation 229 is the total flow time that a molecule spends in the mobile phase, the second term is the total time that a typical molecule is inside beads but unbound to attached ligands, and the third term is the total time that a monovalent molecule is inside a bead and is bound to ligands attached to the bead. Since the differential equations are linear, the probability of being inside a bead "t" units after entering the bead is $\exp(-k_{-1}t)$ so that the mean time for a molecule to stay inside a bead after entering (if it is not chemically bound) is $1/k_{-1}$. Similarly the probability of being bound to a ligand attached to a bead "t" units after becoming bound is $\exp(-k_{-3}t)$ so that the mean-bound time for a molecule to remain bound is $1/k_{-3}$.

The average number of times that a typical molecule moves into and out of a bead if it is not bound chemically is equal to the total unbound time inside the bead divided by the average residence time in the bead for a single entry $(1/k_{-1})$. Therefore

$$\text{average number of entries} = (V_o K_1/F) k_{-1} = V_o k_1/F. \qquad (230)$$

The average number of bindings of a typical monovalent molecule to a ligand-attached bead is the total bound time divided by the average time per binding so that

$$\text{average number of bindings} = \frac{V_pK_3N}{F(1 + K_2L)} k_{-3} = \frac{V_pk_3N}{F(1 + K_2L)} \tag{231}$$

Note that if the flow rate F is very fast, there will be very few entries into beads and very few chemical bindings.

Since our theory involves means, there must be enough occurrences of the processes being averaged so that we can have reasonable confidence in the statistics. Thus the flow rate must be chosen so that the molecules will enter beads and bind to ligands attached to beads many times. We recommended that the flow rate F be slow enough so that the average number of entries and the average number of bindings are each at least 10 and preferably are each over 100. These recommended values are somewhat arbitrary, but it is certainly unreasonable to try to apply the theory if the flow rate is so fast that there are only a few entries or binding events.

For bivalent binding, we find in a similar manner, using Formula 173, that

$$\text{average number of first bindings} = \frac{2V_pk_3N}{F(1 + K_2L)} \tag{232}$$

$$\text{average number of second bindings} = \frac{V_pK_3N\, k_4N}{F(1 + K_2L)^2} \tag{233}$$

For monovalent binding to impenetrable beads we find, using Equation 195, that:

$$\text{average number of bindings} = \frac{V_ok_3N}{(1 + K_2L)} \tag{234}$$

For bivalent binding to impenetrable beads, we obtain using Equation 206 that

$$\text{average number of first bindings} = \frac{2V_ok_3N}{(1 + K_2L)} \tag{235}$$

$$\text{average number of second bindings} = \frac{V_oK_3N\, k_4N}{(1 + K_2L)^2} \tag{236}$$

For reverse-role affinity chromatography, we find using Equation 221 that

$$\text{average number of bindings} = \frac{V_pk_3N}{(1 + K_2I)} \tag{237}$$

G. Estimating Chemical-Reaction Rate Constants From Variances of Elution Profiles

The first theory for measuring chemical-reaction rate constants by affinity chromatography was developed by Denizot and Delaage.[32] As shown in Section III.C.3 their theory is, however, applicable only when the effect of mass-transfer kinetics on the mean and variance of the profile is negligible. Chaiken[5] found that application of the Denizot and Delaage theory to the elution of ribonuclease through a bed of porous beads yields rate constants that are too small by several orders of magnitude.

Here we consider three methods of using affinity chromatography to estimate chemical reaction-rate constants from elution profiles. These methods correspond to the theories developed in Sections III.A to E involving ligands attached to porous beads, ligands attached to impenetrable beads, and macromolecules attached to porous beads. As we explain below, the last two methods have some practical advantages over the first method. Another method for determining chemical-rate constants by counting bound and unbound molecules in an affinity chromatography column is used by Hage et al.[58]

1. Macromolecules Binding to Ligands on Porous Beads

The nonequilibrium theory of affinity chromatography with porous beads includes expressions in Sections III.A and B. for the variance of an elution profile in terms of flow rate, void volume, ligand concentrations, mass-transfer rate constants, and chemical-reaction rate constants. These results depend on the assumptions that the macromolecules and beads are homogeneous, and that the flow rate is large enough so that the effects of diffusion on dispersion or band-broadening are negligible.

A small amount of the monovalent or bivalent macromolecules starts as a thin layer at the top of the bed packed to a height "h" with permeable beads to which some ligands are covalently attached. In general, the variance of the elution profile depends on both the mass-transfer desorption-rate constant and the chemical-reaction dissociation-rate constants. Note that when the chemical-reaction kinetics are fast compared to the mass-transfer kinetics, the variance W_e given by Formula 137 involves the mass-transfer desorption-rate constant k_{-1}, but does not involve the chemical-reaction rate constants k_{-2} and k_{-3}. Unless the macromolecule-ligand reaction rates are slow compared to the mass-transfer kinetic rates, the profile dispersion is predominately due to the nonequilibration of mass transfer. When the chemical kinetics are slow compared to the mass-transfer kinetics, the variance W_e given by formula 147 involves the chemical reaction rate constants k_{-2} and k_{-3}, but does not involve the desorption-rate constant k_{-1}. Thus, a limitation of the method is that it can only be used to estimate a chemical-dissociation rate constant when chemical kinetics are slow compared to sorption-desorption kinetics. We now illustrate this limitation by presenting a sample calculation for monovalent protein.

First assume that the procedure outlined in Section II.D.5 has been followed to determine the mass-transfer rate constants. If the measured quantities are the flow rate $F = 2$ mℓ/min, void volume $V_o = 2$mℓ, mean-elution volume $V_e = 4$mℓ and elution profile variance $W_e = 2$mℓ^2, then the penetrable volume is $V_p = V_e - V_o = 2$mℓ, the equilibrium constant $K_1 = V_p/V_o = 1$ and the desorption-rate constant is

$$k_{-1} = \frac{2FV_p}{W_e} = 0.067/\text{sec} \tag{238}$$

Now we consider the effects of chemical affinity on the elution profile above. Assume that the concentration of ligands on the porous beads available for binding protein is $N = 10^{-4}M$, that there is no free ligand in the column so $L = 0$, that the macromolecule binds only monovalently to the ligands, that the chemical-reaction association-equilibrium constant is $K_3 = 10^5/M$, and that the dissociation-rate constant is $k_{-3} = 10^2/\text{sec}$. Then the predicted mean-elution volume from Equation 131 is

$$V_e = V_o + V_p + V_p K_3 N = 24 \text{ m}\ell \tag{239}$$

Since the mean-elution volume for affinity chromatography is much different from the mean-elution volume for mass transfer alone, an observed mean-elution volume could be used to estimate $K_3 N$, and then K_3 if N were measured.

Since there is no free ligand, we use Formula 151 for the variance W_e and calculate the contributions of each term to the variance.

$$W_e = \left[\frac{2FV_p}{k_{-1}}\right](1 + K_3 N)^2 + \frac{2FV_p K_3 N}{k_{-3}}$$

$$= 2m\ell^2(1 + 10)^2 + 2\left(\frac{2m\ell}{min}\right)\frac{(2m\ell)(10)}{10^2/sec}\left(\frac{1min}{60sec}\right)$$

$$= 242\ m\ell^2 + 0.0133\ m\ell^2 \tag{151}$$

We see that the contribution to the variance of the nonequilibration of the chemical kinetics (0.0133 mℓ^2) is several orders of magnitude smaller than the contribution of the nonequilibration of the mass transfer (242 mℓ^2). Thus an observed variance of an elution profile could not be used to estimate the dissociation constant k_{-3}. In order to obtain accurate estimates of the dissociation-rate constant k_{-3} using affinity chromatography, the experiment must be designed so that the size of the second term in Equation 151 is at least as great as the size of the first term.

There are several possible ways to circumvent this limitation. If the bead diameter is decreased, then the desorption-rate constant k_{-1} should increase since k_{-1} was shown in Equation 113 to be inversely proportional to the square of the bead diameter. Another method would be to use impenetrable beads as described in Sections III.C and D. A third possibility is what we have called reversed-role affinity chromatography in which the macromolecules are attached to the beads and the small-substrate molecules are eluted from the chromatography column. All procedures are subject to the caveats discussed in Section II.A.2.

2. Macromolecules Binding to Ligands on Impenetrable Beads

The obvious advantage of using impenetrable beads instead of porous beads for estimating chemical-reaction rate constants is that there is no mass transfer in and out of the beads so that the nonequilibration of the mass-transfer kinetics cannot dominate the nonequilibration of the chemical kinetics. For monovalent binding of macromolecules to ligands covalently attached to the surface of the impenetrable beads, the variance is given by Formula 196. When there is no free ligand (L = 0), Formula 196 simplifies to $W_e = 2FV_oK_3N/k_{-3}$ so that it might be possible to estimate k_{-3} directly from an observed variance if the other parameter values were known. For bivalent binding to ligands on impenetrable beads when there is no free ligand, the variance is given by Formula 209.

3. Substrate Binding to Macromolecules on Porous Beads

Here we describe how reversed-role affinity chromatography as described in Section III.E can be used to estimate chemical-reaction-rate constants. Since the substrate molecules are much smaller than the macromolecules, they have a larger diffusion constant and move in and out of the beads more easily. Thus the desorption-rate constant k_{-1} is larger for these substrate molecules.[3] Hence the nonequilibration of the mass-transfer kinetics makes a smaller contribution to the elution profile variance so that it may be possible to estimate chemical-reaction-rate constants from observed variances of elution profiles.

We now present a procedure for estimating the dissociation-rate constant k_{-3} and the equilibrium (association) constant K_3 by reversed-role affinity chromatography.

1. Follow the procedure given in Section II.D.5 to determine V_o, V_p, K_1 and k_{-1} when there are no bead-attached macromolecules or when the sites are completely filled by inhibitor.
2. Using preliminary estimates, check to see that Inequality 221, corresponding to slow chemical-reaction kinetics, is satisfied. If the variance in Step 1 was so small that estimates of k_{-1} could not be obtained, then this constitutes experimental evidence that the contribution of nonequilibration of mass transfer to the variance in Step 3 below is negligible so that the formula in Step 4 below can be used.
3. Find values of the mean elution volume V_e and the variance W_e from an observed

elution profile. Since $N^* = N/(1 + K_2I)$, reasonable elution times can be obtained by adjusting the inhibitor concentration I.

4.   Calculate the dissociation-rate constant by using

$$k_{-3} = 2F(V_e - V_o - V_p)/W_e \qquad (239)$$

which was obtained by combining Equations 221 and 222.

5.   If values of the mean-elution volume V_e have been measured for various inhibitor concentrations I, then linear regression can be applied to

$$\frac{V_p}{V_e - V_o - V_p} = \frac{1}{K_3N} + \frac{K_2}{K_3N} I \qquad (240)$$

The ratio of slope to intercept can be used to determine K_2. If N has been measured as described earlier, then K_3 can be estimated from the intercept. Since the error in measuring N may be large, the estimate of the dissociation-rate constant k_{-3} in Step 4 will probably be more accurate than the estimate of K_3 in this step.

6.   Using the estimates obtained above, verify that Inequality 221 corresponding to slow-chemical kinetics is satisfied.

IV. SUMMARY

A. Summary of Chromatography and Affinity Chromatography Formulas

Formulas are presented for the mean and variance of the elution profile for both affinity and nonaffinity chromatography. The numbers of the formulas are their numbers when they were derived in earlier sections. Some of these formulas are simplier than those given in Reference 9 because of the assumption here that the solute molecules equilibrate with the free ligand before the column flow is started. It is assumed in the formulas below that the flow rate is large enough so that diffusion is negligible (see Section II.D.4) and is small enough so that molecules are able to enter the porous beads (see Section III.F.3).

For convenience the notation is reviewed: "h" is the height of the column bed, V_o is the void volume, V_p is the penetrable volume, "u" is the flow velocity, $F = V_o u/h$ is the flow rate, M_e is the mean-elution time, $V_e = FM_e$ is the mean-elution volume, S_e is the variance of the elution profile as a function of time, $W_e = F^2S_e$ is the variance as a function of the volume eluted, k_1 is the sorption-rate constant (for mass transfer into beads), k_{-1} is the desorption-rate constant (for mass transfer out of beads), and $K_1 = k_1/k_{-1}$ is the equilibrium-association constant for mass transfer.

Although they have not been derived here a few formulas for trivalent, tetravalent, and higher order binding are included below. A review of the notation for affinity chromatography is: L is the concentration of free ligand (competitor); k_2 and k_{-2} are the association- and dissociation-rate constants for macromolecules binding to free ligand; $K_2 = k_2/k_{-2}$ is the corresponding equilibrium (association) constant; N is the concentration of available binding sites on ligands attached to beads; k_3, k_{-3}, k_4, k_{-4}, k_5, k_{-5}, k_6, and k_{-6} are the association- and dissociation-rate constants for the first, second, third, and fourth binding of sites on a macromolecule to ligands attached to a bead, respectively; and K_3, K_4, K_5, and, K_6 are the corresponding equilibrium (association) constants, respectively.

1. Means and Variances for Elution Chromatography

The formulas below are for small-zone elution chromatography. The elution-profile peak as an approximation to the mean is discussed in Section II.B.5. Formulas for the third central moments and coefficients of skewness are given in Section II.C.3. Formulas for large-zone experiments when T is the thickness of the initial layer are given in Section II.C.4. Formulas for the mean and variance when the diffusion is not negligible are given in Sections II.D.2 and 3. Formulas for the height equivalent to a theoretical plate and reduced plate height are given in Section II.F.1.

Formulas:

$$M_e = (1 + K_1)\, h/u \tag{39}$$

$$V_e = V_o + V_p \tag{41}$$

$$S_e = 2K_1 h/(uk_{-1}) \tag{40}$$

$$W_e = 2FV_p/k_{-1} \tag{42}$$

2. Profile Means for Binding to Ligands on Porous Beads

The relationship between profile means and peaks is discussed in Section III.F.2. Formulas for trivalent, tetravalent, and higher-order binding are included even though a derivation has not been given here.

Monovalent:

$$M_e = \frac{h}{u}\left[1 + K_1\left(1 + \frac{K_3 N}{1 + K_2 L}\right)\right] \tag{130}$$

$$V_e = V_o + V_p\left(1 + \frac{K_3 N}{1 + K_2 L}\right) \tag{131}$$

Bivalent:

$$M_e = \frac{h}{u}\left[1 + K_1\left(1 + \frac{2K_3 N}{1 + K_2 L} + \frac{K_3 N K_4 N}{(1 + K_2 L)^2}\right)\right] \tag{173}$$

$$V_e = V_o + V_p\left(1 + \frac{2K_3 N}{1 + K_2 L} + \frac{K_3 N K_4 N}{(1 + K_2 L)^2}\right) \tag{174}$$

Trivalent:

$$M_e = \frac{h}{u}\left[1 + K_1\left(1 + \frac{3K_3 N}{1 + K_2 L} + \frac{3K_3 N K_4 N}{(1 + K_2 L)^2} + \frac{K_3 N K_4 N K_5 N}{(1 + K_2 L)^3}\right)\right]$$

$$V_e = V_o + V_p\left(1 + \frac{3K_3 N}{1 + K_2 L} + \frac{3K_3 N K_4 N}{(1 + K_2 L)^2} + \frac{K_3 N K_4 N K_5 N}{(1 + K_2 L)^3}\right)$$

Tetravalent:

$$M_e = \frac{h}{u}\left[1 + K_1\left(1 + \frac{4K_3 N}{1 + K_2 L} + \frac{6K_3 N K_4 N}{(1 + K_2 L)^2} + \frac{4K_3 N K_4 N K_5 N}{(1 + K_2 L)^3} + \frac{K_3 N K_4 N K_5 N K_6 N}{(1 + K_2 L)^4}\right)\right]$$

$$V_e = V_o + V_p\left(1 + \frac{4K_3 N}{1 + K_2 L} + \frac{6K_3 N K_4 N}{(1 + K_2 L)^2} + \frac{4K_3 N K_4 N K_5 N}{(1 + K_2 L)^3} + \frac{K_3 N K_4 N K_5 N K_6 N}{(1 + K_2 L)^4}\right)$$

and in general, for an "n" valent molecule

$$V_e = V_o + V_p \left[1 + \sum_{j=1}^{n} \binom{n}{j} \frac{\prod\limits_{i=3}^{2+j} K_i N^{i-2}}{(1 + K_2 L)^j} \right]$$

where $\binom{n}{j}$ is the binomial coefficient.

3. Profile Variances for Binding to Ligands on Porous Beads

Formulas are given here when the chemical-reaction kinetics are fast compared to the mass-transfer kinetics, or when there is no free-competing ligand. We have derived the Formula 147 for monovalent binding when the chemical kinetics are slow, but it may be too complicated to be useful. We have not obtained the formula for bivalent binding and slow-chemical kinetics, but it would be extremely complicated. Only the formulas for the variance S_e of the elution profile as a function of time are given; however, the variance($W_e = F^2 S_e$) formulas for W_e are the same as those for S_e below except that the factor $2K_1 h/u$ is replaced by $2FV_p$.

Fast Chemical Reactions:

$$S_e = \frac{2K_1 h}{uk_{-1}} \left(1 + \frac{K_3 N}{1 + K_2 L} \right)^2 \qquad\qquad \text{monovalent (136)}$$

$$S_e = \frac{2K_1 h}{uk_{-1}} \left(1 + \frac{2K_3 N}{1 + K_2 L} + \frac{K_3 N K_4 N}{(1 + K_2 L)^2} \right)^2 \qquad\qquad \text{bivalent (180)}$$

$$S_e = \frac{2K_1 h}{uk_{-1}} \left(1 + \frac{3K_3 N}{1 + K_2 L} + \frac{3K_3 N K_4 N}{(1 + K_2 L)^2} + \frac{K_3 N K_4 N K_5 N}{(1 + K_2 L)^3} \right)^2 \qquad\qquad \text{trivalent}$$

$$S_e = \frac{2K_1 h}{uk_{-1}} \left(1 + \frac{4K_3 N}{1 + K_2 L} + \frac{6K_3 N K_4 N}{(1 + K_2 L)^2} + \frac{4K_3 N K_4 N K_5 N}{(1 + K_2 L)^3} + \frac{K_3 N K_4 N K_5 N K_6 N}{(1 + K_2 L)^4} \right)^2 \qquad \text{tetravalent}$$

No free ligand (L = 0):

$$S_e = \frac{2K_1 h}{u} \left[\frac{(1 + K_3 N)^2}{k_{-1}} + \frac{K_3 N}{k_{-3}} \right] \qquad\qquad \text{monovalent (150)}$$

$$S_e = \frac{2K_1 h}{u} \left[\frac{(1 + 2K_3 N + K_3 N K_4 N)^2}{k_{-1}} + \frac{2K_3 N(1 + K_4 N/2)^2}{k_{-3}} + \frac{K_3 N K_4 N}{2k_{-4}} \right] \qquad \text{bivalent (184)}$$

4. Profile Means and Variances for Binding to Ligands on Impenetrable Beads

Monovalent means:

$$M_e = \frac{h}{u} \left(1 + \frac{K_3 N}{1 + K_2 L} \right) \tag{191}$$

$$V_e = V_o \left(1 + \frac{K_3 N}{1 + K_2 L} \right) \tag{195}$$

Bivalent means:

$$M_e = \frac{h}{u}\left[1 + \frac{2K_3N}{1 + K_2L} + \frac{K_3NK_4N}{(1 + K_2L)^2}\right] \qquad (206)$$

$$V_e = V_o\left[1 + \frac{2K_3N}{1 + K_2L} + \frac{K_3NK_4N}{(1 + K_2L)^2}\right] \qquad (207)$$

Monovalent variances with no free ligand:

$$S_e = \frac{2hK_3N}{u\,k_{-3}} \qquad (194)$$

$$W_e = \frac{2FV_oK_3N}{k_{-3}} \qquad (196)$$

Bivalent variances with no free ligand:

$$S_e = \frac{4hK_3N}{uk_{-3}}\left(1 + \frac{K_4N}{2}\right)^2 + \frac{hK_3NK_4N}{uk_{-4}} \qquad (208)$$

$$W_e = \frac{4FV_oK_3N}{k_{-3}}\left(1 + \frac{K_4N}{2}\right)^2 + \frac{FV_oK_3NK_4N}{k_{-4}} \qquad (209)$$

5. Profile Means and Variance for Binding to Macromolecules on Porous Beads

Some of the notation is different for reversed-role affinity chromatography since the macromolecules are attached to the beads and smaller substrate molecules are eluted from the column (see Section III.E). A review of the notation is: S = concentration of substrate, N = concentration of binding sites on bead-attached macromolecules, I = concentration of inhibitor in solvent, and N* = N/(1 + K_2I) is the concentration of free sites available for binding substrate.

Formulas:

$$M_e = \frac{h}{u}\left[1 + K_1(1 + K_3N^*)\right] \qquad (217)$$

$$V_e = V_o + V_p(1 + K_3N^*) \qquad (221)$$

$$S_e = \frac{2K_1h}{u}\left[\frac{(1 + K_3N^*)^2}{k_{-1}} + \frac{K_3N^*}{k_{-3}}\right] \qquad (219)$$

$$W_e = 2FV_p\left[\frac{(1 + K_3N^*)^2}{k_{-1}} + \frac{K_3N^*}{k_{-3}}\right] \qquad (222)$$

REFERENCES

1. **Altgelt, K. H. and Segal, L., Eds.,** *Permeation Chromatography*, Marcel Dekker, New York, 1971.
2. **Ackers, G. K.,** *The Proteins*, 1, 1975, 1.
3. **Yau, W. W., Kirkland, J. J., and Bly, D. D.,** *Modern Size Exclusion Chromatography*, Wiley-Interscience, New York, 1979.
4. **Snyder, L. R. and Kirkland, J. J.,** *Introduction to Modern Liquid Chromatography*, John Wiley & Sons, New York, 1974.
5. **Chaiken, I. M.,** *Anal. Biochem*, 97, 1, 1979.
6. **DeLisi, C. and Hethcote, H. W.,** in *Affinity Chromatography and Related Techniques*, Analytical Chemistry Symposia Series, Vol. 9, Gribnau, T. C. J., Visser, J., and Nivard, R. J. F., Eds., Elsevier, Amsterdam, 1982, 63.
7. **Hethcote, H. W. and DeLisi, C.,** *J. Chrom.*, 240, 269, 1982.
8. **DeLisi, C., Hethcote, H. W., and Brettler, J.,** *J. Chrom.*, 240, 283, 1982.
9. **Hethcote, H. W. and DeLisi, C.,** *J. Chrom.*, 248, 183, 1982.
10. **Hethcote, H. W. and DeLisi, C.,** in *Affinity Chromatography and Biological Recognition*, Chaiken, I. M., Wilchek, M., and Parikh, I., Eds., Academic Press, Orlando, 1983, 119.
11. **DeLisi, C. and Blumenthal, R., Eds.,** *Physical Chemistry of Cell Surface Events and Cellular Regulation*, Elsevier, North Holland, Amsterdam, 1978.
12. **DeLisi, C. and Siraganian, R.,** *J. Immunol.*, 122, 2293, 1979.
13. **Dembo, M. and Goldstein, B.,** *Cell*, 22, 59, 1980.
14. **DeLisi, C.,** *Mol. Immunol.*, 18, 507, 1981.
15. **DeLisi, C. and Crothers, D. M.,** *Biopolymers*, 10, 1809, 1972.
16. **Perelson, A., Goldstein, B., and Rocklin, S.,** *J. Math. Biol.*, 10, 209, 1980.
17. **Eigen, M.,** in *Quantum Statistical Mechanics in the Natural Sciences*, Mintz, S. L. and Wiedermeyer, S. M., Eds., Plenum Press, New York, 1974.
18. **Weigel, F. and DeLisi, C.,** *Am. J. Physiol.*, 243, 475, 1982.
19. **Noyes, R. M.,** *Prog. Reaction Kinetics*, 1, 129, 1961.
20. **Keiser, J.,** *Am. Chem. Soc.*, 86, 5052, 1982.
21. **Berg, H. C. and Purcell, E. M.,** *Biophys. J.*, 20, 193, 1977.
22. **DeLisi, C.,** in *Methods in Enzymology*, V 93, Langone, J. J. and Van Vunakis, H., Eds., Academic Press, N.Y., 1983, 95.
23. **DeLisi, C. and Wiegel, F. W.,** *Proc. Natl. Acad. Sci. U.S.A.*, 78, 5569, 1981.
24. **Wiegel, F. W. and DeLisi, C.,** *Am. J. Physiol.*, 243, R475, 1982.
25. **Schlichting, H.,** *Boundary Layer Theory*, McGraw-Hill, New York, 1968.
26. **Wank, S., DeLisi, C., and Metzger, H.,** *Biochemistry*, 22, 954, 1983.
27. **Lauffenburger, D. and DeLisi, C.,** *Int. Rev. Cytol.*, 84, 269, 1982.
28. **Lyklema, J.,** in *Affinity Chromatography and Related Techniques*, Gribnau, T. J. C. and Nivard, R. J. F., Eds., Elsevier, Amsterdam, 1982, 11.
29. **Van Oss, C. J., Absolum, D. R., and Neumann, A. W.,** in *Affinity Chromatography and Related Techniques*, Gribnau, T. J. C. and Nivard, R. J. F., Eds., Elsevier, Amsterdam, 1982, 29.
30. **London, F.,** The general theory of molecular forces, *Trans. Faraday Soc.*, 33, 8, 1937.
31. **Setlow, R. B. and Poland, E. E.,** *Molecular Biophysics*, Addison-Wesley, Reading, Mass., 1962.
32. **Denizot, F. C. and Delaage, M. A.,** *Proc. Nat. Acad. Sci. U.S.A.*, 72, 4840, 1975.
33. **Giddings, J. C.,** *Dynamics of Chromatography*, Part I, Marcel Dekker, New York, 1965.
34. **Giddings, J. C. and Eyring, H.,** *J. Am. Chem. Soc.*, 59, 416, 1956.
35. **Thomas, H. C.,** *Ann. N.Y. Acad. Sci.*, 49, 161, 1948.
36. **Erdelyi A., et al.,** *Tables of Integral Transforms*, McGraw-Hill, New York, 1954.
37. **Abramowitz, M. and Stegun, I. A., Eds.,** *Handbook of Mathematical Functions*, National Bureau of Standards Applied Mathematics Series No. 55, U.S. Government Printing Office, Washington, D.C., 1964.
38. **Weiss, G.,** in *First Passage Times in Chemical Physics*, Advances in Chemical Physics, Vol. 13, Prigogine, I., Ed., John Wiley & Sons, New York, 1967, 1.
39. **Weiss, G. H.,** *Sep. Sci.*, 5, 51, 1970.
40. **Ouano, A. C.,** in *Advances in Chromatography*, Vol. 15, Giddings, J. C., Ed., Marcel Dekker, New York, 1966, 233.
41. **Gradshteyn, I. S. and Ryzhik, I. M.,** *Tables of Integrals, Series and Products*, Academic Press, New York, 1965.
42. **Grushka, E., Snyder, L. R., and Knox, J. H.,** *J. Chromatogr. Sci.*, 13, 25, 1975.
43. **Horvath, C. and Lin, H. J.,** *J. Chromatogr.*, 126, 401, 1976.
44. **Knox, J. H.,** *J. Chromatogr. Sci.*, 15, 352, 1977.
45. **Brumbaugh, E. E. and Ackers, G. K.,** *J. Biol. Chem.*, 243, 6315, 1968.

46. **van Deemter, J. J., Zuiderweg, F. J., and Klinkenberg, A.,** *Chem. Eng. Sci.,* 5, 271, 1950.
47. **Pecht, I. and Lancet, O.,** in *Chemical Relaxation in Molecular Biology,* Pecht, I. and Ringler R., Eds., Springer, New York, 1977.
48. **Werblin, T. P. and Siskind, G. W.,** *Immunochemistry,* 9, 987, 1972.
49. **Thakur, A. K. and DeLisi, C.,** *Biopolymers,* 17, 1075, 1978.
50. **DeLisi, C.,** *Biopolymers,* 17, 1385, 1978.
51. **Burns, P. F., Campagnoni, C. W., Chaiken, I. M., and Campagoni, A. T.,** *Biochemistry,* 20, 2463, 1981.
52. **Angal, S. and Chaiken, I. M.,** *Biochemistry,* 21, 1574, 1982.
53. **Dunn, B. and Chaiken, I. M.,** *Proc. Natl. Acad. Sci. U.S.A.,* 71, 2382, 1974.
54. **Dunn, B. and Chaiken, I. M.,** *Biochemistry,* 14, 2343, 1975.
55. **Eilat, D. and Chaiken, I. M.,** *Biochemistry,* 18, 790, 1979.
56. **Chaiken, I. M., Eilat, D., and McCormack, W. M.,** *Biochemistry,* 18, 794, 1979.
57. **Inman, J. K.,** in *Affinity Chromatography and Related Techniques,* Gribnau, T. J. C. and Nivard, R. J. F., Eds., Elsevier, Amsterdam, 1982, 29.
58. **Hage, D. S., Walters, R. R., and Hethcote, H. W.,** Split-peak affinity chromatographic studies of the immobilization-dependent adsorption kinetics of protein A, *Anal. Chem.,* 58, 274, 1986.

Chapter 2

ANALYTICAL AFFINITY CHROMATOGRAPHY AND CHARACTERIZATION OF BIOMOLECULAR INTERACTIONS

Harold E. Swaisgood and Irwin M. Chaiken

TABLE OF CONTENTS

I. INTRODUCTION

Macromolecular recognition lies at the root of biology, from molecular transport, enzymatic conversions, and the immune response to multimolecular organization and cellular and multicellular signaling and development. The underlying mechanisms which produce molecular recognition, including the interplay of structural framework, well-defined and specific noncovalent interactions between molecular surfaces, and conformational motility, continue to occupy research efforts. Yet, in this effort and the overall goal to characterize the role of recognition in increasingly complex biological systems, detection and quantitation of macromolecular and cellular interactons, and the methods to do so, remain central to the study of biorecognition in biology and its application in biotechnology.

The use of affinity chromatography as an analytical tool to detect and measure molecular interactions is well-suited to the increasing demand to study these processes in biological systems of variable size and complexity. Noncovalent interactions of small molecules, macromolecules, and multimolecular assemblies can be measured from the elution characteristics of these substances on affinity matrices containing immobilized interactants. In the design of an affinity-chromatography system the central goal is to prepare an immobilized molecular derivative which retains the ability to bind biospecifically to a mobile substance, often a protein or other macromolecule. When this specificity is achieved, powerful separation methods are available for preparative isolation of the desired mobile substance from complex mixtures. Separation can be achieved even when contaminants are closely related in other properties often used as bases for purification, including size and such surface characteristics as charge and hydrophobicity. As opposed to preparative isolation, affinity chromatography can also be a flexible analytical method since the degree of chromatographic retardation of a mobile interactant on an immobilized one is a measure of the equilibrium constant for affinity matrix-mobile component interaction.[1-8] Furthermore, if during elution molecules are included which affect the affinity-matrix binding of eluting mobile components, then the binding parameters for the interaction of such effectors also can be measured; and since for a given affinity matrix, competitors and other effectors can be varied, as can the chemical nature of mobile interactant, binding specificity and dependence on solution conditions such as pH, buffer components, and temperature all can be evaluted. Finally, when binding constants for the immobilized interactant can be compared with those for the soluble analogs, the relatedness of affinity-matrix and solution interactions can be used in designing and evaluating affinity matrices as preparative tools by helping to define the degree of fidelity of an affinity-chromatography matrix as a biospecific reagent.

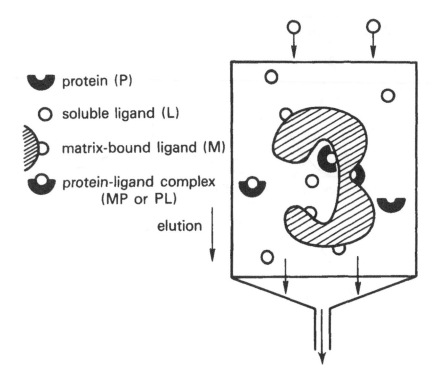

FIGURE 1. Schematic diagram of competitive zonal-elution affinity chromatography depicting the interactions of mobile (P,L) and immobilized (M) interactants. Here, the elution buffer contains a soluble ligand (L) that competes with matrix-bound ligand (M) for binding to the same active site of the mobile interactant (depicted here as a protein, P). (Adapted from Dunn, B. M. and Chaiken, I. M., *Biochemistry*, 14, 2343, 1975. With permission.)

II. FORMULATIONS FOR ZONAL ELUTION QUANTITATIVE AFFINITY CHROMATOGRAPHY

The zonal elution approach of analytical-affinity chromatography was developed essentially in parallel with the continuous elution/frontal analysis approach (Figures 1 and 2). Owing largely to its simplicity of experimental design, the zonal method has become widely applicable for biochemical analysis of molecular interactions when one of the molecular partners can be immobilized with retention of interaction properties. Since the method can be used to analyze very small amounts of the mobile interactant, it has important applicability as a micromethod for biochemical analysis of a vast array of biologically active molecules being discovered and isolated but available in only small amounts (for example, as immunoreactive or radiolabeled species).

Experimentally, zonal-elution affinity chromatography of a mobile interacting molecule on a matrix of immobilized interactant involves three basic steps.

1. A zone of small volume containing the mobile component is applied onto a column equilibrated with appropriate buffer with or without effector
2. Elution is continued with equilibration buffer ($+/-$ effector)
3. The elution profile of the initially applied mobile component is monitored

A. Monovalent Systems
The generalized scheme in Figure 1 defines a set of competing monovalent-binding reactions,

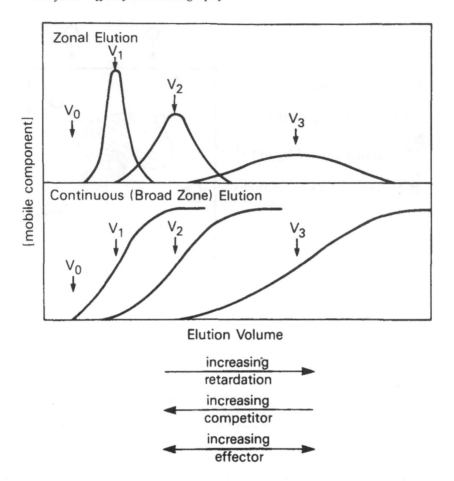

FIGURE 2.   Schematic representation of expected elution behavior in (top) zonal-elution and (bottom) continuous-elution quantitative affinity chromatography. Using the case of Figure 1, zonal-quantitative affinity chromatography is performed by introducing a small-volume aliquot of protein (mobile component) solution in the presence of varying concentration of soluble ligand to an affinity column of immobilized ligand (the column is pre-equilibrated with buffer with appropriate concentration of soluble ligand) and eluting with buffer-containing soluble ligand (but not mobile component) until the zone of protein is detected in the eluate. The elution volume (V) of the protein zone and the peak width decrease as the concentration of soluble competing ligand increases. Continuous (broad zone) elution is performed as above except that a relatively large volume of the solution of protein (mobile component, with or without competing ligand) is introduced continuously to the affinity column until the eluted solution shows a plateau of mobile component concentration equal to that of the solution applied to the column. In both cases, values of V can be used to determine quantitative binding properties of protein to immobilized ligand as well as to soluble competing ligand. Elutions also can be carried out in the presence of effectors which either may enhance (increase V) or suppress (decrease V) binding of mobile and immobilized components; in such cases, binding properties of effectors also can be determined.

$$L + P \overset{K_{L/P}}{\longleftrightarrow} LP$$

$$M + P \overset{K_{M/P}}{\longleftrightarrow} MP \qquad (1)$$

where P is the mobile interactant, M is the matrix-immobilized interactant, L is a soluble component that competes with M for binding to P; LP and MP are noncovalent complexes formed between mobile and immobilized species, respectively; and K_{LP} and K_{MP} are dissociation constants for LP and MP, respectively. For the purposes of most of the present discussion, P is defined as a macromolecule, such as a protein, and L and M are small molecular ligands that bind to the same site on P. The equations to be discussed are also valid for all cases which follow the mechanism given by Equation 1, including when both P and M are macromolecules.

Partitioning of P is defined as

$$\sigma_P = \frac{[MP] + [P^\alpha] + [LP^\alpha]}{[P^\beta] + [LP^\beta]} \tag{2}$$

where α and β refer to stationary and mobile (bulk) phase concentrations, respectively. When $[M]_T = O$ (the concentration of immobilized ligand is O), the partitioning of P is given by

$$\sigma_{O,P} = \frac{[P^\alpha] + [LP^\alpha]}{[P^\beta] + [LP^\beta]} \tag{3}$$

an expression which would take into account nonspecific interaction with the matrix and inaccessible pore volume. The partitioning of P in the presence of immobilized interactant becomes

$$\sigma_P = \frac{[MP]}{[P^\beta] + [LP^\beta]} + \sigma_{O,P} = \frac{[MP]\,\sigma_{O,P}}{[P^\alpha] + [LP^\alpha]} + \sigma_{O,P} \tag{4}$$

For chromatography in general,

$$V_i = V_m + \sigma_i\,V_s \tag{5}$$

where V_i is the elution volume for species i, σ_i is its partition coefficient, V_m is the mobile-(bulk) phase volume, and V_s is the stationary-phase volume. When no specific interaction occurs (e.g., $[M]_T = O$), the unretarded elution volume, V_o, is given by

$$V_{o,i} = V_m + \sigma_{o,i}\,V_s \tag{6}$$

Using Equation 4, the chromatographic equation for interactant P is given by

$$V - V_m = \left(\frac{[MP]}{[P^\alpha] + [LP^\alpha]} + 1 \right) V_s$$

or

$$\frac{V - V_o}{V_o - V_m} = \frac{[MP]}{[P^\alpha] + [LP^\alpha]} \tag{7}$$

If nonspecific partitioning does not occur and the total pore volume is accessible to all interactants, then $[P^\alpha] = [P^\beta] = [P]$ and $[L^\alpha] = [L^\beta] = [L]$. Substitution of the expressions for molecular equilibria from Equation 1 into Equation 7 yields

$$\frac{V - V_o}{V_o - V_m} = \frac{[M] \, K_{L/P}}{K_{M/P}(K_{L/P} + [L])} \tag{8}$$

The total concentration of immobilized interactant is given by

$$[M]_T = [M] \left(1 + \frac{[P]}{K_{M/P}} \right) \tag{9}$$

Consequently

$$\frac{V - V_o}{V_o - V_m} = \frac{[M]_T}{K_{M/P} + [P]} \cdot \frac{K_{L/P}}{K_{L/P} + [L]}$$

which, when inverted, gives

$$\frac{V_o - V_m}{V - V_o} = \frac{K_{M/P}}{[M]_T} + \frac{K_{M/P} \, [L]}{K_{L/P} \, [M]_T} + \left(1 + \frac{[L]}{K_{L/P}} \right) \frac{[P]}{[M]_T} \tag{10}$$

We note from this expression that if $[P] \ll [M]_T$, as is often the case for zonal elution with porous matrices, and if $[P]$ is sufficiently small so that $[L] \gg [LP]$, the equation becomes

$$\frac{V_o - V_m}{V - V_o} = \frac{K_{M/P}}{[M]_T} + \frac{K_{M/P} \, [L]_T}{K_{L/P} \, [M]_T}$$

or

$$\frac{1}{V - V_o} = \frac{K_{M/P}}{[M]_T \, (V_o - V_m)} + \frac{K_{M/P} \, [L]_T}{K_{L/P} \, [M]_T \, (V_o - V_m)} \tag{11}$$

Equation 11 was used initially[1] to evaluate quantitative affinity chromatographic data by relating the variation of V, the experimentally measured elution volume for mobile inter-actant, to the total concentrations of immobilized and soluble competing ligands, $[M]_T$ and $[L]_T$, respectively. The values for $[M]_T$, V_o, and V_m are constants defined for a particular matrix and physical arrangement of the column. $[L]_T$ also is experimentally measurable. Thus, the dissociation constants $K_{M/P}$ and $K_{L/P}$ can be obtained directly from experimental chromatographic data. For a specific affinity matrix of fixed $[M]_T$, a series of elutions is carried out in which $[L]_T$ is varied. The elution volumes of zones of mobile macromolecule are determined at the various values of $[L]_T$. From a plot of $1/(V - V_o)$ or $V_o - V_m)/(V - V_o)$ vs. $[L]_T$, values for $K_{M/P}$ and $K_{L/P}$ can be calculated as defined by Equation 11. Thus, $K_{L/P}$ can be derived from the ratio of slope/ordinate intercept and $K_{M/P}$ from the intercept directly. Values for $K_{L/P}$ and $K_{M/P}$ can also be calculated nongraphically by linear least-squares regression analysis of the elution data.

In some instances, it may be desirable or necessary to perform zonal elutions in the absence of competing soluble ligand ($[L]_T = O$). In this situation Equation 11 simplifies to

$$\frac{V_o - V_m}{V - V_o} = \frac{K_{M/P}}{[M]_T} \quad \text{or} \quad \frac{1}{V - V_o} = \frac{K_{M/P}}{[M]_T \, (V_o - V_m)} \tag{12}$$

Thus, for the same matrix of fixed $[M]_T$ that is used for competitive zonal elutions, a value for $K_{M/P}$ can be calculated independently from a single elution profile.

Equations 11 and 12 above are valid for quantitative purposes when the concentration of protein is small so that [L]»[LP] and therefore $[L]_T = [L]$. However, if this condition does not hold, the last term in Equation 10 cannot be neglected. Since the concentration of P is given by $[P] = [P]_T/(1 + [L]/K_{L/P})$, Equation 10 becomes

$$\frac{V_o - V_m}{V - V_o} = \frac{K_{M/P} + [P]_T}{[M]_T} + \frac{K_{M/P} [L]}{K_{L/P} [M]_T} \tag{13}$$

which is equivalent to an equation first derived by Nichol et al.[2] for this case. Since the value of $[P]_T$ is not definable (it changes continuously during elution) in zonal-elution chromatography, it is necessary to use the alternative procedure of continuous-elution (or large-zone) chromatography at constant $[P]_T$ and frontal analysis to evaluate the elution volume.[2] However, continuous-elution (or large-zone) quantitative-affinity chromatography can be more time-consuming than zonal elution and impractical when the mobile interactant is available in only limited amounts. Consequently, for many interacting systems of interest, zonal-elution analysis is likely to be the method of choice.

B. Bivalent Systems

Zonal-elution chromatography also can be applied to evaluate bivalent-binding systems, including the type in Equation 14.

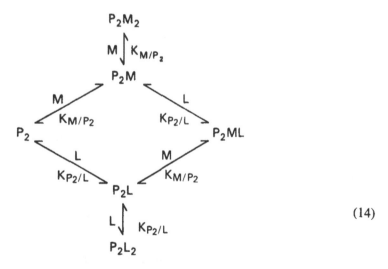

$$(14)$$

In a manner similar to that for monovalent-binding systems and assuming all K_{M/P_2}'s and $K_{P_2/L}$'s are equal, this scheme leads[9,10] to the formulation

$$\frac{1}{V - V_o} = \frac{1 + 2\left(\dfrac{[L]}{K_{P_2/L}}\right) + \left(\dfrac{[L]}{K_{P_2/L}}\right)^2}{(V_o - V_m)\left[2\left(\dfrac{[M]_T}{K_{M/P_2}}\right) + \left(\dfrac{[M]_T}{K_{M/P_2}}\right)^2 + 2\left(\dfrac{[L][M]_T}{K_{P_2/L} K_{M/P_2}}\right)\right]} \tag{15}$$

for sufficiently low concentrations of P_2. This expression allows microscopic-dissociation constants, $K_{P_2/L}$ and K_{M/P_2}, to be evaluated for a bivalent-binding species, P_2 by measuring V at varying [L]. However, unlike the case for monovalent-binding systems for these conditions, the variation of $1/(V - V_o)$ with [L] for bivalent systems is nonlinear. Thus, the values of $K_{P_2/L}$ and K_{M/P_2} are derived from competitive-elution data by nonlinear least-squares regression analysis. Despite this difference, the experimental protocol for collecting data for a bivalent-binding system is the same as for a monovalent system.

When zonal elutions are carried out without soluble ligand present ([L] = O), Equation 15 simplifies to

$$V - V_o = (V_o - V_m) \left[2\left(\frac{[M]_T}{K_{M/P_2}}\right) + \left(\frac{[M]_T}{K_{M/P_2}}\right)^2 \right] \tag{16}$$

Thus, at [L] = O, the value of K_{M/P_2} can be calculated directly from the data obtained in a single zonal elution.

C. Multiple Equilibria with Immobilized-Protein Systems

As a quantitative tool, affinity chromatography using a protein as the immobilized inter-actant offers some unique possibilities for characterization of protein-protein and protein-ligand interactions. This approach, for example, could be used to characterize enzyme-subunit or enzyme-enzyme interactions as affected by effector ligands. In this discussion, we will specifically consider neurophysin (M)-neurophysin (P) and neurophysin (M or P)-vasopressin (L) interactions.[11-14] The equilibria occurring with the immobilized interactant are

$$(17)$$

Similar equilibria occur in the mobile phase when both P and L are present. It should be noted that for this case M is immobilized P.

Analogous to Equation 2, the partitioning of protein is given by

$$\sigma_P = \frac{[MP] + [MLP] + [MPL] + [MPL_2] + [P^\alpha] + [PL^\alpha] + 2[P_2^\alpha] + 2[PPL^\alpha] + 2[PLP^\alpha] + 2[P_2L_2^\alpha]}{[P^\beta] + [PL^\beta] + 2[P_2^\beta] + 2[PPL^\beta] + 2[PLP^\beta] + 2[P_2L_2^\beta]}$$

Defining $\sigma_{o,p}$ analogous to Equation 3, this becomes

$$\sigma_P = \frac{([MP] + [MLP] + [MPL] + [MPL_2])\, \sigma_{o,p}}{[P^\alpha]_T} + \sigma_{o,p} \tag{18}$$

where $[P^\alpha]_T$ is the total concentration of protein, not matrix-bound, in the stationary phase on a monomer-molecular weight basis. Thus, using Equations 5 and 6, the chromatography expression is

$$V - V_o = \frac{([MP] + [MLP] + [MPL] + [MPL_2])(V_o - V_m)}{[P^\beta]_T} \tag{19}$$

with the assumption that $[P^\beta]_T = [P^\alpha]_T$. Substituting expressions for the various equilibria and assuming that the concentrations of soluble protein, $[P]$, and ligand, $[L]$, are the same in the mobile phase and in the pore volume, we obtain

$$\frac{V - V_o}{V_o - V_m} = \frac{\dfrac{[M][P]}{K_{M/P}}\left(1 + \dfrac{[L]}{K_{M^*P/L}} + \dfrac{[L]}{K_{MP^*/L}} + \dfrac{[L]^2}{K_{MPL/L}K_{MP^*/L}}\right)}{[P]_T} \tag{20}$$

The total concentration of immobilized protein molecules also can be related to the concentration of dissociated species and equilibrium constants

$$[M]_T = [M]\left(1 + \frac{[L]}{K_{M/L}}\right) + \frac{[M][P]}{K_{M/P}}\left(1 + \frac{[L]}{K_{MP^*/L}} + \frac{[L]}{K_{M^*P/L}} + \frac{[L]^2}{K_{MPL/L}K_{MP^*/L}}\right) \tag{21}$$

Defining the last term in brackets in Equations 21 and 20 as Q, these expressions become

$$[M]_T = [M]\left(1 + \frac{[L]}{K_{M/L}} + \frac{[P]\,Q}{K_{M/P}}\right) \tag{22}$$

and

$$\frac{V - V_o}{V_o - V_m} = \frac{[M]_T\,[P]\,Q}{K_{M/P}\,[P]_T\left(1 + \dfrac{[L]}{K_{M/L}} + \dfrac{[P]\,Q}{K_{M/P}}\right)} \tag{23}$$

Taking the inverse of Equation 23 gives

$$\frac{V_o - V_m}{V - V_o}\,([M]_T) = \frac{K_{M/P}}{Q}\frac{[P]_T}{[P]} + \frac{K_{M/P}}{K_{M/L}}\frac{[L][P]_T}{Q\,[P]} + [P]_T \tag{24}$$

Equation 24 is analogous to Equation 10 for a monovalent system.

The concentration of unliganded-protein monomer can be calculated if equilibrium constants for soluble species are available. Thus, it can be shown by conservation of mass that

$$[P]^2 + \frac{K_{P/P}}{2Q}\left(1 + \frac{[L]}{K_{P/L}}\right)[P] - \frac{K_{P/P}}{2\,Q}[P]_T = 0$$

and solution of this quadratic equation yields

$$[P] = \frac{K_{P/P}}{4Q}\left(1 + \frac{[L]}{K_{P/L}}\right)\left[\left(1 + \frac{8Q[P]_T}{K_{P/P}\left(1 + \dfrac{[L]}{K_{P/L}}\right)^2}\right)^{1/2} - 1\right] \tag{25}$$

Both Equations 24 and 25 are complicated functions of the concentration of unbound ligand, L, involving all of the immobilized liganded-protein dissociation constants; consequently, it is unlikely that, without simplification, the individual constants can be extracted from chromatographic data. However, there are two experimentally accessible special cases of interest.

Specific case No. 1 — Chromatography of the protein in the absence of ligand. When $[L] = O$, $Q = 1$; thus Equations 24 and 25 simplify to

$$\frac{V_o - V_m}{V - V_o} ([M]_T) = (K_{M/P}) \left(\frac{[P]_T}{[P]}\right) + [P]_T = [P]_T \left(\frac{K_{M/P}}{[P]} + 1\right) \quad (26)$$

and

$$[P] = \frac{K_{P/P}}{4} \left[\left(1 + \frac{8[P]_T}{K_{P/P}}\right)^{1/2} - 1\right] \quad (27)$$

Since $[P]_T = [P] + 2[P]^2/K_{P/P}$ when $[L] = O$, combining this result with Equations 25 to 27 gives

$$\frac{V_o - V_m}{V - V_o} = \frac{K_{M/P}}{2[M]_T} + \left[\frac{K_{M/P}}{2} \left(1 + \frac{8[P]_T}{K_{P/P}}\right)^{1/2} + [P]_T\right] \frac{1}{[M]_T} \quad (28)$$

Independently determined values for $[M]_T$ and $K_{P/P}$ allow determination of $K_{M/P}$ by nonlinear least-squares fitting of zonal-elution data as a function of the total soluble-protein concentration, $[P]_T$. More significantly, when $[P]_T \ll K_{P/P}$ and $K_{M/P}$, the quantity in the brackets becomes $K_{M/P}/2$. Consequently,

$$\frac{V_o - V_m}{V - V_o} = \frac{K_{M/P}}{[M]_T} \text{ (when } [P]_T << K_{P/P} \text{ and } K_{M/P}) \quad (29)$$

Specific case No. 2 — Chromatography of the protein in the presence of saturating ligand. When L is saturating, the only matrix-bound species present at significant concentrations are ML and MPL_2. Therefore, the relationship for elution volume analogous to Equation 7 (with all previous assumptions) is

$$\frac{V - V_o}{V_o - V_m} = \frac{[MPL_2]}{[PL^\alpha] + 2[P_2L_2^\alpha]} = \frac{[ML][PL]}{K_{ML/PL} [P]_T} \quad (30)$$

Likewise, analogous to Equation 9, the total concentration of immobilized molecules, $[M]_T$, is given by

$$[M]_T = [ML] \left(1 + \frac{[PL]}{K_{ML/PL}}\right) \quad (31)$$

and Equation 25 becomes

$$[PL] = \frac{K_{PL/PL}}{4} \left[\left(1 + \frac{8[P]_T}{K_{PL/PL}}\right)^{1/2} - 1\right] \quad (32)$$

Combination and rearrangement of Equations 30 to 32 yields

$$\frac{V_o - V_m}{V - V_o} = \frac{K_{ML/PL}}{2\,[M]_T} + \left[\frac{K_{ML/PL}}{2}\left(1 + \frac{8[P]_T}{K_{PL/PL}}\right)^{1/2} + [P]_T\right]\frac{1}{[M]_T} \qquad (33)$$

This relationship has exactly the same form as Equation 28 except that the dissociation constants pertain to the fully liganded dimers, both immobilized and soluble. Likewise, when $[P]_T \ll K_{ML/PL}$ and $K_{PL/PL}$ Equation 33 reduces to

$$\frac{V_o - V_m}{V - V_o} = \frac{K_{ML/PL}}{[M]_T} \quad (\text{when } [P]_T \ll K_{ML/PL} \text{ and } K_{PL/PL}) \qquad (34)$$

As a result, the dissociation constants for immobilized unliganded and fully liganded dimers are readily obtained from zonal-elution quantitative affinity chromatography if sufficiently sensitive methods are used for detection of the eluting soluble protein to allow experiments to be carried out at low $[P]_T$.

Chromatography of the ligand molecule — Because of the equilibria between monovalent and bivalent species, chromatography of liganded molecules in the presence of protein capable of association with the immobilized species is not simply the inverse of the monovalent or bivalent systems discussed above. Thus, the partition coefficient is more complex, the relationship analogous to Equation 4 being

$$\sigma_L = \frac{([ML] + [MLP] + [MPL] + 2[MPL_2])\,\sigma_{O,L}}{[L^\alpha] + [PL^\alpha] + [PPL^\alpha] + [PLP^\alpha] + 2[P_2L_2^\alpha]} + \sigma_{O,L} \qquad (35)$$

The corresponding expression for chromatography of the ligand molecule in the presence of associating protein molecules thus becomes

$$\frac{V - V_o}{V_o - V_m} = \frac{\dfrac{[M][L]}{K_{M/L}} + \dfrac{[M][P]}{K_{M/P}}\left(\dfrac{[L]}{K_{M\cdot P/L}} + \dfrac{[L]}{K_{MP\cdot/L}} + \dfrac{2[L]^2}{K_{MPL/L}\,K_{MP\cdot/L}}\right)}{[L]_T} \qquad (36)$$

Substitution for [M] using Equation 22 and rearrangement gives

$$\frac{V - V_o}{V_o - V_m}\left(\frac{[L]_T}{[M]_T}\right) = \frac{\dfrac{[L]}{K_{M/L}} + \dfrac{[P]}{K_{M/P}}\left(\dfrac{[L]^2}{K_{MP\cdot/L}\,K_{MPL/L}} + Q - 1\right)}{1 + \dfrac{[L]}{K_{M/L}} + \dfrac{[P]}{K_{M/P}}\,Q} \qquad (37)$$

Thus, the zonal-elution parameters are a complex function of individual equilibrium constants and the concentrations of dissociated species, P and L. However, if protein P is not included in the elution buffer the equation simplifies to

$$\frac{V_o - V_m}{V - V_o} = \frac{K_{M/L}}{[M]_T} + \frac{[L]_T}{[M]_T} \qquad (38)$$

where $[L]_T$ is the total-ligand concentration in the mobile phase. From this relationship we note that *the elution volume is linearly related to the ligand concentration in the mobile phase*. However, for chromatography using porous matrices $[L^\beta]_T$ is considerably less than the initial concentration of ligand in the zone applied, $[L°]_T$, due to zone dispersion, and $[M]_T$ is relatively large. Consequently, it is relatively easy, using sensitive methods for detection of the eluting ligand, to obtain data in the region where

$$\frac{V_o - V_m}{V - V_o} \cong K_{M/L}/[M]_T \tag{39}$$

D. Relationships for Chromatography on Nonporous Beads

There are situations for which it is desirable to perform quantitative affinity chromatography using nonporous matrices. For this case the partition coefficient is defined in a manner analogous to that for porous matrices, except that the void volume replaces the stationary phase or pore volume in the relationship, i.e.

$$\sigma \equiv \frac{Q_s}{V_o C}$$

where Q_s is the amount of solute adsorbed to the matrix and C is the solute concentration in the mobile phase. The chromatographic expression is then

$$V_i = V_m + \sigma_i V_o \tag{40}$$

When $[M]_T = 0$, $\sigma_o V_o = V_o - V_m$, so if there is no nonspecific adsorption, i.e., $V_o = V_m$, then $\sigma_o V_o = 0$. Hence, $V - V_o = \sigma V_o$ or

$$\frac{V - V_o}{V_o} = \sigma \tag{41}$$

Equation 41 is analogous to Equations 7, 19, and 30 for porous matrices; all of the previous equations have the same form except $V_o/(V - V_o)$ replaces $(V_o - V_m)/(V - V_o)$.

However, for nonporous matrices $[M]_T$ usually will be quite small, consequently, the terms containing its reciprocal in the previous equations cannot be neglected. Thus, we have from Equation 28

$$\frac{V_o}{V - V_o} = \frac{K_{M/P}}{[M]_T} + \frac{[P]_T}{[M]_T} \quad \text{when} \quad [P]_T \ll K_{P/P} \tag{42}$$

from Equation 33

$$\frac{V_o}{V - V_o} = \frac{K_{ML/PL}}{[M]_T} + \frac{[P]_T}{[M]_T} \quad \text{when} \quad [P]_T \ll K_{PL/PL} \tag{43}$$

and from Equation 38

$$\frac{V_o}{V - V_o} = \frac{K_{M/L}}{[M]_T} + \frac{[L]_T}{[M]_T} \tag{44}$$

In these equations $[P]_T$ and $[L]_T$ are the total concentrations of P and L in the mobile phase. Unfortunately, in zonal chromatography this value is constantly changing so the concentration at a specific time is a function of the original concentration applied, $f[P]_o$ or $f[L]_o$. Since this function is unknown, the ratio of intercept/slope cannot be used to obtain $K_{M/L}$, $K_{M/P}$, or $K_{ML/PL}$. However, at low concentrations the relationship, with respect to the initial concentrations, appears to be essentially linear, so that extrapolation to give the intercept is accurate and values for the dissociation constants can be obtained when $[M]_T$ is known (see below).

E. General Multiple Equilibria between Soluble Ligands and Immobilized Macromolecules

In many cases of interest, the immobilized species may have multiple binding sites for a soluble ligand. For example, if there are "n" sites of class "i" for binding of L to M, the equilibria are then

$$M + L \leftrightarrow ML \; ; \; [ML] = [M][L]/K_1$$

$$ML + L \leftrightarrow ML_2 \; ; \; [ML_2] = [M][L]^2/K_1K_2$$

$$ML_{n-1} + L \leftrightarrow ML_n \; ; \; [ML_n] = [M][L]^n/K_1K_2 \cdots K_n \tag{45}$$

The partitioning of L is then given by

$$\sigma_L = \frac{[L^\alpha] + [ML] + 2[ML_2] \cdots + n[ML_n]}{[L^\beta]}$$

$$= \frac{[M]\left(\dfrac{[L^\alpha]}{K_1} + \dfrac{2[L^\alpha]^2}{K_1K_2} \cdots + \dfrac{n[L^\alpha]^n}{K_1K_2 \cdots K_n}\right) + [L^\alpha]}{[L^\beta]}$$

Since $\sigma_{O,L} = [L^\alpha]/[L^\beta]$, we have

$$\sigma_L = [M]\left(\frac{1}{K_1} + 2\frac{[L^\alpha]}{K_1K_2} \cdots + \frac{n[L^\alpha]^{n-1}}{K_1K_2 \cdots K_n}\right)\sigma_{O,L} + \sigma_{O,L} \tag{46}$$

Using Equations 5 and 6, the chromatography expression is

$$\frac{V - V_o}{V_o - V_m} = [M]\left(\frac{1}{K_1} + 2\frac{[L^\alpha]}{K_1K_2} \cdots + \frac{n[L^\alpha]^{n-1}}{K_1K_2 \cdots K_n}\right) \tag{47}$$

Since

$$[M]_T = [M]\left(1 + \frac{[L]}{K_1} + \frac{[L]^2}{K_1K_2} \cdots + \frac{[L]^n}{K_1K_2 \cdots K_n}\right) \tag{48}$$

and assuming $[L^\alpha] = [L^\beta] = [L]$, we have

$$\frac{V - V_o}{V_o - V_m} = \frac{[M]_T\left(\dfrac{[L]}{K_1} + 2\dfrac{[L]^2}{K_1K_2} \cdots + \dfrac{n[L]^n}{K_1K_2 \cdots K_n}\right)}{\left(1 + \dfrac{[L]}{K_1} + \dfrac{[L]^2}{K_1K_2} \cdots + \dfrac{n[L]^n}{K_1K_2 \cdots K_n}\right)[L]} \tag{49}$$

which is the same as Equation 8 derived by Lagercrantz et al.[15] However, the ratio of the terms in brackets is the moles of L bound per mole of M, i.e., $\bar{v}$, as defined by Tanford[16]

$$\bar{v} = \cfrac{\cfrac{[L]}{K_1} + \cfrac{2[L]^2}{K_1 K_2} \cdots + \cfrac{n[L]^n}{K_1 K_2 \cdots K_n}}{1 + \cfrac{[L]}{K_1} + \cfrac{[L]^2}{K_1 K_2} \cdots + \cfrac{[L]^n}{K_1 K_2 \cdots K_n}}$$

$$= \sum_{i=1}^{N} \frac{n_i \, k_i \, [L]}{1 + k_i \, [L]} \tag{50}$$

where k_i is the intrinsic association constant for the "ith" class of sites and there are N classes of sites (in these expressions we have defined $1/K_1 \equiv nk_i$, $1/K_2 \equiv (n - 1)k_2/2$, $1/K_i \equiv (n + 1 - i)k_i/i$). Using this relationship, Equation 49 becomes

$$\frac{V - V_o}{V_o - V_m} = \frac{[M]_T}{[L]} \bar{v} = [M]_T \sum_{i=1}^{N} \frac{n_i \, k_i}{1 + k_i \, [L]} \tag{51}$$

As noted by Lagercrantz et al.,[15] if chromatography is performed under conditions such that $[L]/K_i$ and $k_i \, [L] \ll 1$, then

$$\frac{V - V_o}{V_o - V_m} = \frac{[M]_T}{K_1} = [M]_T \sum_{i=1}^{N} n_i \, k_i \tag{52}$$

Thus, K_1 can be obtained by performing zonal chromatography at sufficiently low concentrations of the soluble ligand. This dissociation constant for complexes containing 1 ligand molecule can be a useful parameter for comparison of the binding of various ligands to the macromolecule.

III. EXPERIMENTAL EXAMPLES OF ZONAL ANALYTICAL AFFINITY CHROMATOGRAPHY — MONOVALENT- AND BIVALENT-INTERACTING COMPLEXES

Several protein-ligand and protein-protein interacting systems have been studied by zonal-elution quantitative affinity chromatography. Representative examples are given below, showing the types of chromatographic results typically obtained for monovalent- (e.g., Staphylococcal nuclease/nucleotide and ribonuclease/nucleotide) and bivalent- (e.g., TEPC 15 IgA bivalent monomer/phosphorylcholine) interacting complexes.

A. Staphylococcal Nuclease/Nucleotide and Ribonuclease/Nucleotide

The elutions in Figure 3 were obtained when zones of Staphylococcal nuclease were eluted on pdTpAP-Sepharose at different concentrations of the soluble competitor, pdTpAP.[3] Similar data for ribonuclease on UDP-Sepharose are shown in Figure 4. The linearity of the $1/(V - V_o)$ vs [L] plots in these figures reflects the 1:1 nature of the enzyme/nucleotide interactions in both cases and allows calculation of the dissociation constants $K_{P/L}$ and $K_{M/P}$ using the monovalent model, Equation 11. The calculated constants are given in Table 1. The table also contains results for other competitors, connoting the potential of the method to straightforwardly determine equilibrium constants for a series of competitors and therein to characterize specificity properties of the interactions. For example, the analytical approach of affinity chromatography has been used[25] to show that [4-Fluoro-His 12]RNase, a ribonuclease engineered to be inactive with retention of conformation, can bind active-site ligands, including substrates (Figure 5).

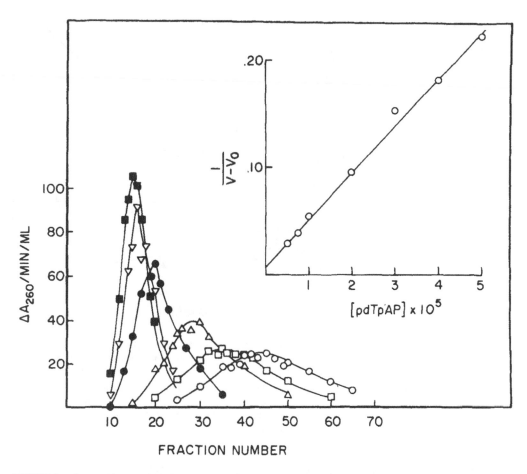

FIGURE 3. Competitive zonal-elution affinity chromatography for a monovalent binding system. Zones containing equal amounts of *Staphylococcal nuclease* were applied to a pdTpAP-Sepharose column (0.9 × 15 cm, [M] ≈ 0.05 m*M*) equilibrated with 0.1 *M* ammonium acetate, pH 5.7, containing soluble competitive ligand (pdTpAP) at concentrations of 0.5 × 10^{-5} *M* (○), 0.785 × 10^{-5} *M* (□), 1.0 × 10^{-5} *M* (△), 2 × 10^{-5} *M* (●), 3 × 10^{-5} *M* (▽), and 4.0 × 10^{-5} *M* (■) and eluted with these same buffers. Elutions were carried out at room temperature. Inset: data from the main figure (and additional data at 5 × 10^{-5} *M* [pdTpAP] omitted from the main figure to avoid clutter) are plotted as $1/(V - V_o)$ versus [pdTpAP]. Dissociation constants, $K_{P/L}$ and $K_{M/P}$ were calculated from the linear plot using Equation 11 and are listed in Table 1. (Adapted from Dunn, B. M. and Chaiken, I. M., *Biochemistry*, 14, 2343, 1975. With permission.)

B. Bivalent Immunoglobulin A (TEPC 15 IgA Monomer)/Phosphorylcholine

When TEPC 15 IgA, which has been reduced and alkylated selectively to yield bivalent monomers, was eluted on immobilized phosphorylcholine at 5 × 10^{-5} *M* immobilized-ligand concentration (determined by functional capacity), competitive elutions with soluble phosphorylcholine could be achieved but the response of $1/(V - V_o)$ to the soluble competitor concentration was nonlinear.[9] This behavior, shown in Figure 6, is consistent with the potential for bivalent binding of IgA monomer to the affinity matrix. As a result, the interaction parameters $K_{P/L}$ and $K_{M/P}$ were evaluated using the bivalent model of Equation 14. These values are given in Table 1. When bivalent IgA is eluted on phosphorylcholine-Sepharose at 9 × 10^{-6} *M* immobilized-ligand concentration, variation of elution volume with soluble phosphorylcholine conforms, as shown in Figure 7, to the monovalent model. Here, the reduced density of matrix-binding sites eliminates any significant chance for simultaneous binding of protein to two matrix sites. Thus, the monovalent as well as bivalent models can be used to calculate $K_{P/L}$ and $K_{M/P}$ values as given in Table 1.

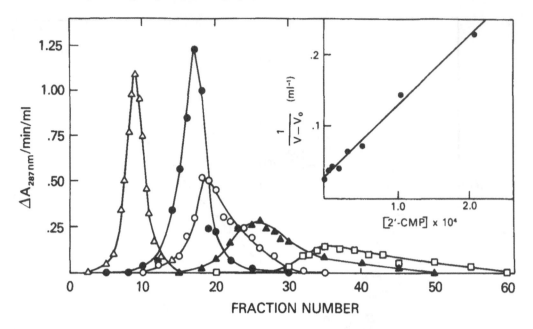

FIGURE 4.   Competitive elution profiles of 2.8 mg of RNase-A on Sepharose-APpUp, in 0.4 M ammonium acetate, pH 5.2, and ambient temperature, containing various molar concentrations of 2'-CMP as follows: (△) 2.1 × 10^{-4}, (●) 5.2 × 10^{-5}, (○) 3.2 × 10^{-5}, (▲) 2.0 × 10^{-5}, and (□) 5.7 × 10^{-6}. $\Delta A_{287\,nm}$/min/mℓ is enzymic activity against cytidine 2':3'-monophosphate. Inset: plot of data according to Equation 11 with the solid line representing the best fit calculated by nonlinear least squares. For data obtained at 1.0 × 10^{-4}, 1.0 × 10^{-5}, and 0 M 2'-CMP, the derived points are included in the inset plot but the corresponding elution profiles are omitted from the main figure for clarity. (Adapted from Chaiken, I. M. and Taylor, H. C., *J. Biol. Chem.*, 72, 629, 1976. With permission.)

While IgA-monomer elution on the higher concentration-affinity matrix (5 × 10^{-5} M) leads to curvilinear $1/(V - V_o)$ vs [L] profiles, elution of monovalent IgA-derived Fab fragments on the same matrix shows linear behavior, as seen in Figure 8. This verifies the interpretation that the curvilinearity with IgA monomers indeed results from incipient bivalent binding to the affinity matrix when the binding sites are close together (high-density substitution).

IV. CONTINUOUS ELUTION ANALYTICAL-AFFINITY CHROMATOGRAPHY: FORMULATIONS AND EXPERIMENTAL APPLICATIONS

A. General Description

An important property of zonal analytical-affinity chromatography is that the concentration of mobile interactant, [P] or [L], continuously changes as the zone passes through the matrix bed. Thus, explicit terms involving [P] or [L] are not included in expressions relating elution volumes and dissociation constants (see equations above, for example). Neglecting [P] or [L] is valid when this parameter is small with respect to $[M]_T$ and to the dissociation constant ($K_{M/P}$ or $K_{M/L}$) for binding of mobile and immobilized interactants. The use of low [P] or [L] in zonal-elution analysis is analogous to the use of low [enzyme] in steady state-enzyme kinetic analysis.[28] Carrying out zonal-elution experiments at low [P] or [L] normally is easily achieved and, indeed, desirable when analyzing biomolecular species of finite availability. It has been shown experimentally (see for example Reference 17) that when the amount of mobile interactant applied to the affinity matrix is low relative to $K_{M/L}$, the dependence of calculated dissociation constants on the amount of mobile interactant used is negligible.

Table 1
COMPARISON OF DISSOCIATION CONSTANTS OBTAINED BY QUANTITATIVE AFFINITY CHROMATOGRAPHY WITH THOSE OBTAINED BY OTHER METHODS — A SELECTED LIST

Protein	Ligand	Chromatographic K_d (M)		K_d(M) by other methods		
		$K_{P/L}{}^a$	$K_{M/P}{}^b$	Value	Method	Ref.
Staphylococcal nuclease	pdTp^c	2.5×10^{-6}		2.5×10^{-6} 5.9×10^{-6}	Equilibrium dialysis Kinetics as enzyme inhibitor	3
	pdTpAP^d	2.3×10^{-6}	1.1×10^{-6e}	2.5×10^{-6}	Kinetics as enzyme inhibitor	
	NPpdTp^f	1.1×10^{-5}		6.3×10^{-6}	Kinetics as enzyme inhibitor	
Bovine pancreatic ribonuclease	2'-CMP^g	1.6×10^{-5}		9.7×10^{-6}	Kinetics as enzyme inhibitor	17
	APpUp^h		9.3×10^{-6e}	1.7×10^{-5}	Kinetics as enzyme inhibitor	
Immunoglobulin A (TEPC 15) monovalent FAB	Phosphoryl choline	$1.5\text{—}3.3 \times 10^{-6}$	$3.9\text{—}4.2 \times 10^{-6i}$	3.0×10^{-6}	Equilibrium dialysis	9,10
Immunoglobulin A (TEPC 15) divalent monomer	Phosphoryl choline	$1.2\text{—}1.5 \times 10^{-6}$	$2.7\text{—}4.8 \times 10^{-6i,j}$ $1.2 \times 10^{-6,k}$ $1.3 \times 10^{-7,l}$	2.0×10^{-6}	Equilibrium dialysis	
Bovine lactate dehydrogenase H$_4$ (heart)	NADH	3.8×10^{-7}		3.9×10^{-7}	Fluorescence titration	18
M$_4$ (muscle)	NADH	1.1×10^{-6}		2.0×10^{-6}	Quenching of protein fluorescence	
Rabbit muscle lactate dehydrogenase	NADH	1.1×10^{-5}		1.0×10^{-5}	Frontal gel filtration chromatography	19
Trypsin	N$^\alpha$-Acetyl-Gly-Gly-Arg	5.9×10^{-4}	1.3×10^{-4m}	4.7×10^{-4n}	Kinetics as enzymic inhibitor	4,20

Table 1 (continued)
COMPARISON OF DISSOCIATION CONSTANTS OBTAINED BY QUANTITATIVE AFFINITY CHROMATOGRAPHY WITH THOSE OBTAINED BY OTHER METHODS — A SELECTED LIST

Protein	Ligand	Chromatographic K_d (M)		K_d(M) by other methods		Ref.
		$K_{P/L}{}^a$	$K_{M/P}{}^b$	Value	Method	
Trypsin	β-Aminobenzamidine	1×10^{-5}	1.2×10^{-6}	8.2×10^{-6}	Kinetics as enzymic inhibitor	21
	Benzamidine	$1.6—3.9 \times 10^{-5}$		1.8×10^{-5}	Kinetics as enzymic inhibitor	
α-Chymotrypsin			2.4×10^{-5o}	2.4×10^{-5}	Ultracentrifugation	6
Bovine glutamate dehydrogenase	Perphenazine	2×10^{-6}	6×10^{-6}	2×10^{-5p}	Kinetics as enzymic inhibitor	22
	Chlorpromazine	4×10^{-6}		3.2×10^{-5p}	Kinetics as enzymic inhibitor	
	Trifluperidol	22×10^{-6}		14×10^{-5}	Kinetics as enzymic inhibitor	
Rabbit muscle lactate dehydrogenase	Cibacron Blue		0.3×10^{-6q}	$0.1—0.5 \times 10^{-6}$	Enzyme catalysis and spectrophotometric titration	23
	NADH	3.4×10^{-6}		0.5×10^{-6}	Enzyme catalysis and spectrophotometric titration	
Rabbit muscle lactate dehydrogenase	Cibacron Blue		$\sim 0.5 \times 10^{-6r}$	$0.1—0.5 \times 10^{-6}$	Enzyme catalysis and spectrophotometric titration	24

a K_d for soluble ligand as determined by competitive elution.
b K_d for immobilized ligand.
c   Thymidine-3',5'-diphosphate.
d   Thymidine-3'-(p-aminophenylphosphate)-5'-phosphate.
e   For ligand immobilized to agarose through aminophenyl moiety.
f   Thymidine-3'-phosphate-5'-(p-nitrophenylphosphate).

g   Cytidine-2'-monophosphate.

h   Uridine-5'-(4-aminophenylphosphate)-2'(3')-phosphate.

i   For ligand immobilized to agarose through a glycyl(azophenyl)tyrosyl arm on the phosphate moiety.

j   Microscopic K_{MP} determined using Equation 15.

k   Functional K_{MP} determined at $[M]_T = 9 \times 10^{-6}$ M using Equation 11.

l   Functional K_{MP} determined at $[M]_T = 5 \times 10^{-5}$ M using Equation 11.

m   For ligand immobilized to agarose through α-amino of Gly-Gly-Arg.

n   Determined at pH 6.0 vs. pH 6.2 and 6.0 for corresponding $K_{P/L}$ and K_{MP}, respectively.

o   For chymotrypsin immobilized at pH 8.

p   Relative hierarchy of chromatographically derived K_d is the same as that for kinetically derived values; quantitative differences in kinetic vs. chromatographic values observed consistently, suggesting contribution of matrix.

q   Corrected for functional capacity.

r   Value obtained after correcting for observed degree of bivalency.

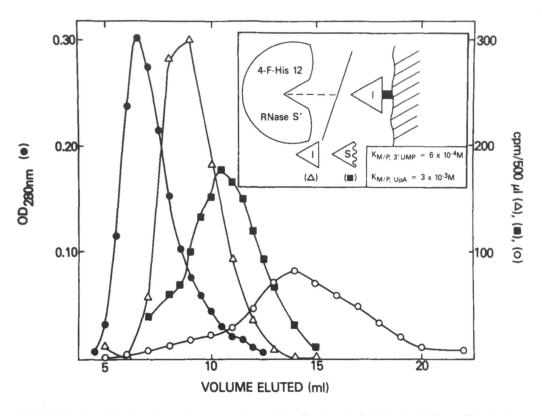

FIGURE 5.   Zonal-elution affinity chromatographic analysis of semisynthetically-derived [4-fluoro His 12, des 16-20]-SRNase S, an active site analog of RNase A.[25-27] Single elution profiles are shown of bovine serum albumin (2.8 mg) and [4-F-His 12, des 16-20]SRNase S (0.1 mg) on Sepharose-APpUp, in 0.4 M ammonium acetate, pH 5.2, 23°C and active site ligand concentrations as indicated: (●) bovine serum albumin in buffer alone, (△) [4-F-His 12, des 16-20]SRNase S in 2.2 × 10^{-3} M 3'-UMP, (■) [4-F-His 12, des 16-20]SRNase S in 2.0 × 10^{-4} M UpA, and (○) [4-F-His 12, des 16-20]SRNase S in buffer alone. (Adapted from Taylor, H. C. and Chaiken, I. M., *J. Biol. Chem.*, 252, 6991, 1977. With permission.)

The limitations of undefined [P] or [L] can be circumvented by continuous elution of the solute or by application of a zone sufficiently large that a boundary is established in the mobile phase between the solvent and the solute reaching a plateau concentration equal to the concentration in the applied solution. An analysis of the boundary then can be performed analogous to that for sedimentation velocity and free-electrophoresis experiments.

The position of such a boundary whose movement corresponds to that for solute molecules in the plateau region is the first moment of the concentration vs. volume curve.[29] Experimentally, this is given by

$$\overline{V} = \frac{\sum_i V_i \, \Delta \, [P]_i}{[P]_o} = na - \frac{a \sum_{i=1}^{n} [P]_i}{[P]_o} \tag{53}$$

where $[P]_o$ is the initial solute concentration, "a" is the volume of each fraction, and "n" is a position in the plateau region.[29,30] The position $\overline{V}$ is that of a hypothetical infinitely sharp boundary and corresponds to the position where $[P] = [P]_o / 2$ if the gradient curve for the boundary is Gaussian.

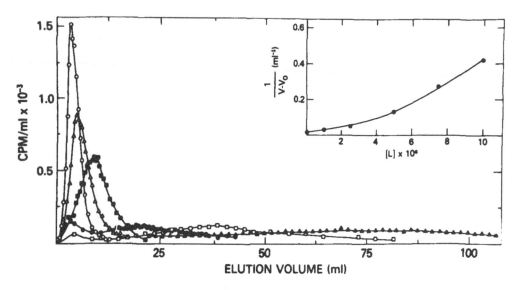

FIGURE 6. Competitive zonal-elution affinity chromatography for a bivalent binding system. Zones (100 μℓ) of [¹⁴C]IgA monomer (bivalent binding) were applied to high-density phosphorylcholine-Sepharose (7 × 25 mm, [M] = 5 × 10⁻⁵ M) equilibrated in PBS (with 1 mg/mℓ bovine serum albumin) and containing soluble phosphorylcholine at the following concentrations (M); 0 (▲), 1 × 10⁻⁶ (□), 2.5 × 10⁻⁶ (●), 5.0 × 10⁻⁶ (■), 7.5 × 10⁻⁶ (△), 1.0 × 10⁻⁶ (○). Elutions were carried out at room temperature with buffer containing the indicated amount of soluble competitive phosphorylcholine. Inset: elution data are plotted as $1/(V - V_o)$ vs. [phosphorylcholine]. Dissociation constants, $K_{P/L}$ and $K_{M/P}$, were calculated from the nonlinear plot using Equation 15 and are shown in Table 1. (From Eilat, D. and Chaiken, I. M., *Biochemistry*, 18, 790, 1979, and Chaiken, I. M., Eilat, D., and McCormick, W. M., *Biochemistry*, 18, 794, 1979. With permission.)

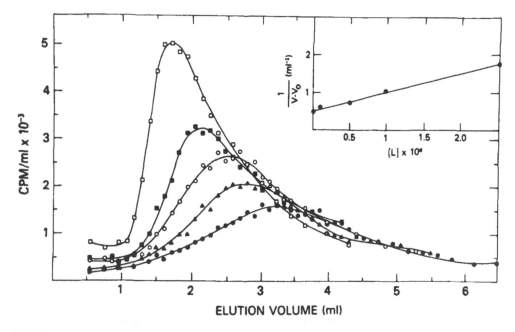

FIGURE 7. Competitive-zonal elutions of bivalent IgA monomers on low-density phosphorylcholine-Sepharose. Zones of [¹⁴C]IgA monomer were applied to low-density phosphorylcholine-Sepharose (7 × 25 mm, [M] = 9 × 10⁻⁶ M) equilibrated and eluted as described for the high-density phosphorylcholine-Sepharose column in Figure 6 except that the following molar concentrations of soluble phosphorylcholine were used: 0 (●), 1 × 10⁻⁷ (▲), 5 × 10⁻⁷ (●), 1 × 10⁻⁶ (■), and 2.5 × 10⁻⁶ (□). Inset: variation of elution volumes (V) with concentration of free phosphorylcholine (L) plotted as $1/(V - V_o)$ vs. [L]. (From Eilat, D. and Chaiken, I. M., *Biochemistry*, 18, 790, 1979, and Chaiken, I. M., Eilat, D., and McCormick, W. M., *Biochemistry*, 18, 794, 1979. With permission.)

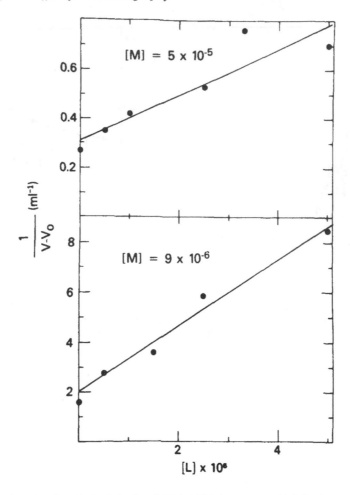

FIGURE 8.   Elution behavior of [³H]IgA Fab fragments on high-density (upper) and low-density (lower) phosphorylcholine-Sepharose columns, plotted as the variation of $1/(V - V_o)$ with soluble competing phosphorylcholine concentration. Chromatographic conditions for elutions of zones of [³H]IgA Fab fragments are as described in Figure 6 for IgA monomer. Dissociation constants $K_{P/L}$ and $K_{M/P}$ were calculated from the linear plots using Equation 11 and are shown in Table 1. (Adapted from Eilat, D. and Chaiken, I. M., *Biochemistry*, 18, 790, 1979, and Chaiken, I. M., Eilat, D., and McCormick, W. M., *Biochemistry*, 18, 794, 1979. With permission.)

All of the equations derived in the section on zonal elution apply for continuous- or large-zone chromatography using the elution volume defined by the first moment of the boundary to compute the chromatographic parameters corresponding to the initial concentration of the solute applied. The relevant expressions are Equations 13, 28, 33, 38, 42 to 44, and 51.

B. Trypsin-Ligand Interactions

Frontal analysis of continuous-elution boundaries has been used to characterize the interaction of the inhibitor benzamidine and β-trypsin. The interaction was studied using Gly-Gly-Arg Sepharose and both β-trypsin and inhibitor in the mobile phase[30] as well as the inverse system of immobilized β-trypsin with the inhibitor in the mobile phase.[31]

Chromatography of trypsin in the presence of inhibitor conforms to the monovalent system discussed previously and described by Equation 10. In the absence of inhibitor, this expression can be rearranged to give

$$K_{M/P} = \frac{V_o - V_m}{\bar{V} - V_o} [M]_T - [P]_o \tag{54}$$

Since $V_o = V_m + V_s$ when $\sigma_{o,p} = 1$ (i.e., all of the stationary phase is penetrable), $V_s - V_o = V_m$ and $V_s [M]_T = M_T$, the total amount of immobilized ligand. Thus, we have

$$K_{M/P} = \frac{M_T}{\bar{V} - V_o} - [P]_o \tag{55}$$

which corresponds to the authors' Equation 2.[31] Furthermore, in the limit as $[P] \rightarrow 0$, Equation 55 becomes

$$K_{M/P} = \frac{M_T}{\bar{V}_{lim} - V_o} \tag{56}$$

Also, with the above conditions, Equation 11, after rearrangement, becomes

$$\frac{1}{1 + \dfrac{[L]_T}{K_{L/P}}} = \frac{K_{M/P}}{M_T} (\bar{V}_i - V_o) \tag{57}$$

which upon introduction of Equation 56 yields

$$\frac{1}{1 + \dfrac{[L]_T}{K_{L/P}}} = \frac{\bar{V}_i - V_o}{\bar{V}_{lim} - V_o} \tag{58}$$

This equation can be rearranged to a linear form for plotting

$$\frac{[L]_T}{K_{L/P}} = \frac{\bar{V}_{lim} - \bar{V}_i}{\bar{V}_i - V_o}$$

or

$$\bar{V}_i = V_o + K_{L/P} \frac{\bar{V}_{lim} - \bar{V}_i}{[L]_T} \tag{59}$$

which corresponds to their Equation 5. Plotting chromatographic elution data for β-trypsin in the presence of benzamidine according to this equation (Figure 9) yielded a dissociation constant, $K_{L/P}$, of 14 μM which was in excellent agreement with the inhibition constant obtained kinetically, 15 μM.[30] The chromatographic method is applicable under conditions where there is little enzymic activity and thus kinetic methods cannot be used. Consequently, the binding of inhibitor could be investigated as a function of pH between 3 and 8 and of temperature between 4 and 25°C.[30]

In some respects, use of the inverse system, i.e., immobilized trypsin, is more convenient and capable of yielding more information, such as the effect of solute concentration on binding affinity. Thus, Equation 55 can be used directly and rearranged to give

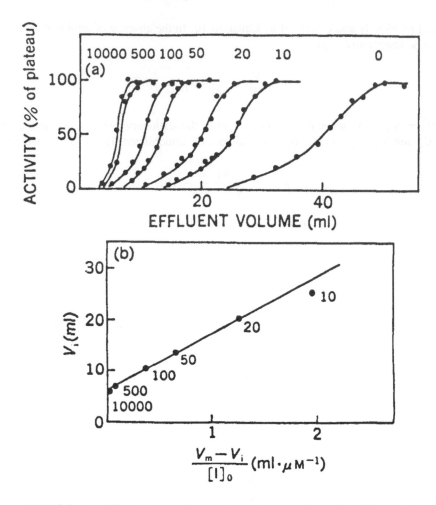

FIGURE 9.   (a) Effect of benzamidine on the continuous elution profile of β-trypsin. A column of GlyGlyArg-Sepharose (0.6 × 10 cm) was equilibrated and run with 0.1 M Tris-maleate buffer containing 0.02 M CaCl$_2$, pH 6.0. The temperature was 4°C. The concentration of enzyme was 0.2 μM. (b) Plot of V$_i$ against (V$_m$ − V$_i$)/[I]$_o$ (our Equation 59). The benzamidine concentrations are indicated in μM units. (From Kasai, K. and Ishii, S., *J. Biochem.*, 84, 1051, 1978. With permission.) (In their notation V$_m$ corresponds to our V$_{lim}$.)

$$[P]_o(\overline{V} - V_o) = \frac{M_T [P]_o}{K_{M/P} + [P]_o} \qquad (60)$$

Noting the analogy of this relationship to that for enzyme kinetics, one can use the same statistical methods which have been developed to analyze kinetic data in order to evaluate chromatographic data. For example, the weighted least-squares regression analysis of Wilkinson[32] and the direct linear-plot analysis developed by Eisenthal and Cornish-Bowden[33] were used by Kasai and Ishii[31] to evaluate the binding of inhibitors to immobilized trypsin. The data can also be plotted according to various linearized forms of the equation such as the double-reciprocal form[31] or the more statistically accurate Hanes form ([S]/v vs. [S])

$$\frac{1}{\overline{V} - V_o} = \frac{K_{M/P}}{M_T} + \frac{1}{M_T} [P]_o \qquad (61)$$

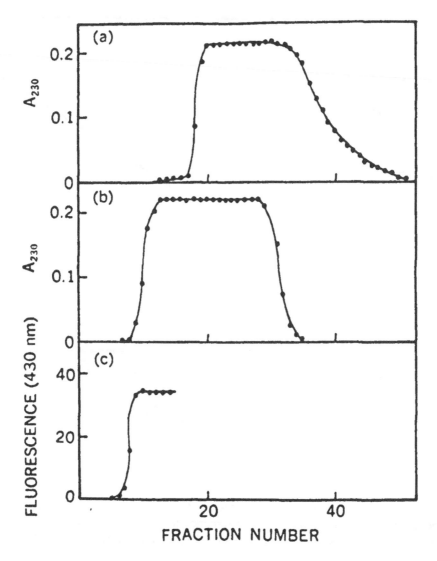

FIGURE 10.   Continuous-elution chromatography of benzamidine on β-trypsin Sepharose. The column (0.9 × 10.5 cm) was equilibrated and run with 0.05 M acetate buffer containing 0.01 M CaCl$_2$, pH 6.0. The temperature was 4°C. Fractions of 1.025 mℓ were collected. (a) 21.5 mℓ of 19.9 μM benzamidine was applied. (b) The same experiment as (a) except for the addition of leupeptin (50 μg/mℓ) which prevents binding of benzamidine. (c) 10 μM glycine. (From Kasai, K. and Ishii, S., *J. Biochem.*, 84, 1061, 1978. With permission.)

(Note that for this inverse system where M is immobilized trypsin, the solute P represents the inhibitor.)

A typical elution profile for benzamidine on immobilized β-trypsin is shown in Figure 10. Analysis of all the chromatographic data according to Equation 60, using either the weighted regression method of Wilkinson or the direct linear plot method of Eisenthal and Cornish-Bowden, gave values for M_T, the total amount of active immobilized trypsin in the column, and $K_{M/P}$, the dissociation constant for the trypsin-inhibitor complex. The two methods of analysis gave excellent internal agreement and the values obtained were essentially the same as those from the inverse system and from kinetic analysis. The values of M_T and $K_{M/P}$ obtained by Kasai and Ishii[31] were used to construct the plot, shown in Figure 11, according to Equation 60 to emphasize the analogy with enzyme-kinetic data.

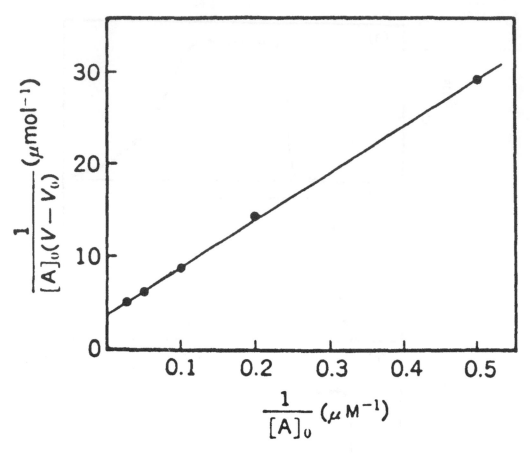

FIGURE 11.   Plot of $1/[A]_o(V - V_o)$ vs. $1/[A]_o$ (double-reciprocal form of Equation 60) for benzamidine elution on β-trypsin Sepharose. (From Kasai, K. and Ishii, S., *J. Biochem.*, 84, 1061, 1978. With permission.)

The power of this technique is apparent when it is recognized that immobilized proteins can be quite stable and that once M_T is determined for a column, the dissociation constant for any protein-ligand complex can be obtained by simply measuring V and V_o for that ligand.

C. Binding of Organic Acids to Bovine-Serum Albumin

The association of various organic acids with serum albumin is an excellent example of the case of multiple-independent binding sites for a ligand on the surface of a protein. Classically, such interactions have been analyzed by Scatchard plots of equilibrium-dialysis data.[34] However, as shown by Equation 51, the moles of ligand bound per mole of protein, $\bar{v}$, can be obtained directly from frontal analysis of affinity-chromatographic data. For example, a Scatchard plot of data obtained by chromatography of salicylic acid on immobilized-serum albumin is shown in Figure 12. Also, it is noted from Equation 52 that the dissociation constant for the first ligand-molecule bound is related to these Scatchard site-binding constants

$$\frac{1}{K_1} = \sum_{i=1}^{N} n_i k_i$$

Lagercrantz et al.,[15] observed that data for salicylic acid obtained by zonal- and continuous-elution chromatography were in agreement with this relationship, supporting the validity of these methods for analysis.

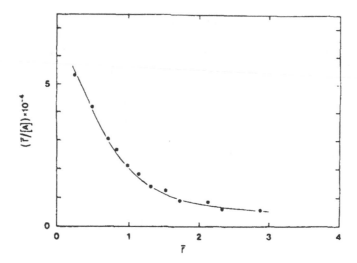

FIGURE 12. Scatchard diagram of the system [^{14}C]salicylic acid-salicylic acid. The curve was constructed using $k_1 = 6.42 \times 10^4 \, M^{-1}$ for one site and seven sites with $k_2 = 7.32 \times 10^2 \, M^{-1}$ and expressions derived from our Equation 51. Note that $\bar{r}$ in this figure corresponds to $\bar{\nu}$ in our notation. Experimental data points: ●. (From Lagercrantz, C., Larsson, T., and Karlsson, H., *Anal. Biochem.*, 99, 352, 1979. With permission.)

D. Chromatography of Multivalent Components

Nichol and co-workers[35,36] have treated chromatography of a multivalent-mobile component in the presence of a univalent ligand which competes with matrix-bound ligand for sites on the multivalent component. Assuming that there is no interaction between sites on the mobile component and that binding is governed by k_{MP}, the single-site binding constant of soluble P to the matrix, the authors showed that this association constant is given by

$$k_{MP} = \frac{(1 + k_{PL} [L]) [1 - ([P^\beta]_T/[P]_T)^{1/f}]}{\left(\dfrac{[P^\beta]_T}{[P]_T}\right)^{1/f} ([M]_T - f[P]_T [1 - ([P^\beta]_T/[P]_T)^{1/f}])} \qquad (62)$$

where k_{PL} is the association constant for binding of soluble P and soluble ligand, "f" is the valency of the multivalent component, and $[P^\beta]_T$ and $[P]_T$ are the total concentrations of multivalent component in the mobile phase and in the total system, respectively. (The notation in their expression has been changed to that employed in the present discussion.)

The above expression can be written in terms of chromatographic parameters, V and V_o, used in this discussion. If the amount of P applied to the chromatographic column is Q_T, this amount will be distributed between the phases according to

$$Q_T = [P^\beta]_T V_m + Q_s = [P^\beta]_T V_m + \sigma[P^\beta]_T V_s$$

assuming that the concentration of soluble P in the accessible volume of the stationary phase is equal to the concentration in the mobile phase. Thus

$$Q_T = [P^\beta]_T (V_m + \sigma V_s) = [P^\beta]_T \overline{V} \qquad (63)$$

If the same amount of P is applied to a column without immobilized ligand,

$$Q_T = [P^\beta]_T (V_m + \sigma_o V_s) = [P^\beta]_T V_o \tag{64}$$

where V_o is the total volume in the column accessible to P. Thus, if the total concentration of P is defined in terms of the total accessible volume, we obtain

$$[P]_T = \frac{Q_T}{V_o} = [P^\beta]_T \frac{\overline{V}}{V_o} \tag{65}$$

Substitution of this result into Equation 62 gives

$$k_{MP} = \frac{(1 + k_{PL} [L]) \left[1 - \left(\frac{V_o}{\overline{V}} \right)^{1/f} \right]}{\left(\frac{V_o}{\overline{V}} \right)^{1/f} \left([M]_T - f[P^\beta]_T \frac{\overline{V}}{V_o} \left[1 - \left(\frac{V_o}{\overline{V}} \right)^{1/f} \right] \right)} \tag{66}$$

If "f" is unknown, a series of experiments must be performed in the absence of ligand L using varying values of $[M]_T$ and $[P^\beta]_o$. The assumed constancy of k_{MP} then allows both "f" and k_{MP} to be determined. Additional experiments with varying [L] allow k_{PL} to be determined using conditions where $[L] = [L]_T$, i.e., $[L]_T \gg [P]_T$.*

V. KINETICS OF THE INTERACTION BETWEEN SOLUBLE MOLECULES AND IMMOBILIZED SPECIES: FORMULATIONS AND EXPERIMENTAL EXAMPLES

A. Theory

In addition to equilibrium parameters, which can be derived from the mean-elution volume, the spreading of the zone for zonal-elution chromatography also contains information related to the kinetics of association and dissociation of the mobile component-immobilized component interaction. There have been two theoretical treatments of such zone dispersion which have provided equations for evaluation of the association- and dissociation-rate constants.

The first approach, taken by Denizot and Delaage,[38] was an extension of the statistical theory of chromatography formulated by Giddings and Eyring. This treatment resulted in the following relationships for the association-rate constant, k_a, and the dissociation-rate constant, k_d

---

* Nichol et al.,[35] first reported a relationship for k_{MP} in terms of chromatographic parameters. However, their derivation misdefined the total concentration of component P in the column, $[P]_T$ as equal to $[P^\beta]_T (1 + K_{AV})$, where the partition coefficient K_{AV} was defined by

$$K_{AV} = \frac{\overline{V} - V_m}{V_t - V_m}$$

(our terminology, except V_t = total column volume). More correctly, as incorporated in the present paper in the derivation of Equation 66

$$[P]_T = \frac{[P^\beta]_T V_m + [P^\alpha]_T V_s}{V_m + V_s}$$

where $[P^\beta]_T$ and $[P^\alpha]_T$ are the total concentrations of P in the mobile and stationary phases, respectively, and the quantity $(V_m + V_s)$ is the total liquid volume in the column. Since $[P^\alpha]_T = K_{AV} [P^\beta]_T$, we see that $[P]_T = [P^\beta]_T (1 + K_{AV})(V_s/2V_s)$ if $V_m = V_s$. Thus, the relationship $[P]_T = [P^\beta]_T (1 + K_{AV})$ implies that $V_m = V_s$ and that P is distributed only in a volume equal to that of the stationary phase. Hogg and Winzor[37] recently presented a corrected relationship of k_{MP} equivalent to Equation 66 of the present discussion.

$$k_a = \frac{2\,E(t_o)\,[E(t') - E(t_o)]^2}{\sigma'^2\,E^2(t_o) - \sigma_o^2\,E^2(t')} \tag{67}$$

$$k_d = \frac{2\,E^2(t_o)\,[E(t') - E(t_o)]}{\sigma'^2\,E^2(t_o) - \sigma_o^2\,E^2(t')} \tag{68}$$

where $E(t_o)$ and $E(t')$ are the peak positions in time units for unretarded molecules and molecules which bind, respectively, and σ_o and σ' are their respective standard deviations, also in time units. The substance used to determine the characteristics of unretarded molecules should have similar hydrodynamic properties to those of the molecule whose binding is being studied. One approach is to elute a small amount of radiolabeled form in the presence of excess unlabeled form such that all of the binding sites on the matrix are saturated.

A second treatment, developed by Hethcote and DeLisi[39-42] and also analogous to the random-walk model of chromatography (Giddings and Eyring), does not assume local mass-transfer equilibrium. In the absence of a competing soluble molecule that is similar to the immobilized molecule, all of their expressions for chromatography of molecules which interact monovalently with the immobilized molecule reduce to our Equation 39 with V equal to the mean-elution volume. For chromatography with porous beads, when the flow rate is sufficiently large so that the effects of diffusion are negligible and when the chemical reactions are slow compared to mass-transfer rates, it is possible to calculate the chemical-dissociation rate, k_d, from the variance of the elution profile. Their treatment also provides expressions to test these two criteria; for example, the contribution of chemical-reaction kinetics to the variance dominates that caused by mass transfer when the following inequality holds

$$k_d(1 + [M]_T/K_{M/L})^2/([M]_T/K_{M/L}) << k_{-1} \tag{69}$$

where k_{-1} is the mass-transfer desorption rate constant. The value of the latter constant can be calculated from the variance, W_e, of the profile for an unretarded molecule. Thus, if all of the pore volume is accessible as assumed in this chapter

$$k_{-1} = 2F\,V_s/[W_e - V_m^2 T^2/(12\,h^2)] \tag{70}$$

where F is the volumetric flow rate, T is the sample thickness, and "h" is the column length. When the above conditions are obtained, the chemical-dissociation rate can be calculated from the variance of the elution profile of a homogeneous solute by use of the expression

$$k_d = \frac{2F(V - V_m - V_s)}{W_e} = \frac{2F(V - V_o)}{W_e} \tag{71}$$

Various approaches have been suggested to minimize the contribution of mass-transfer kinetics to the dispersion including the use of small diameter beads and, for the case of a macromolecule-small molecule interaction, immobilization of the macromolecule so that the mobile phase contains the more rapidly diffusing small molecule. The most obvious means for eliminating the effects of mass transfer is to use nonporous beads, thus eliminating the pore volume entirely. Indeed, Hethcote and DeLisi have argued that the expressions derived by Denizot and Delaage are applicable only for the case of nonporous beads. Hence, for nonporous beads, k_d can be calculated from the profile variance using Equations 68 or 71 noting that $V_s = 0$ so $V_m = V_o$.

B. Experimental Tests with Vasopressin and Neurophysin

The kinetics of the dimerization of bovine neurophysin II (BNP II) and of its binding to the neuropeptide Arg^8-vasopressin (AVP) have been examined by analytical high-performance affinity chromatography with BNP II immobilized on nonporous-glass beads. The chromatography of ^{3}H-AVP on this affinity matrix was examined as a function of AVP concentration by addition of varying amounts of unlabeled AVP to a standard amount of ^{3}H-AVP. A typical elution profile, shown in Figure 13A, illustrates the symmetry of the profiles obtained in this manner. The solid line in this figure represents a fit to a Gaussian equation from which both the elution volume, V, and the profile variance, W_e, were obtained. These data can be used to calculate the dissociation-rate constant by either the Hethcote-DeLisi method or the Denizot-Delaage method as illustrated in Table 2. Essentially the same value is obtained for the dissociation-rate constant with either method when nonporous beads are used, as was predicted by Hethcote and DeLisi. However, this value is smaller by several orders of magnitude than that obtained when both molecules are in solution. This result may be due to the probability of rebinding to another site when these sites are part of a contiguous surface. Nevertheless, these observations may be of biological relevance since a similar situation would exist when sites are part of a membrane such as for cell receptors or other types of membrane recognition sites.

The BNP II dimer dissociation-rate constant was also evaluated by chromatography of ^{125}I-BNP II on a BNP II-nonporous-glass bead column. Following experimental procedures similar to those for chromatography of ^{3}H-AVP, elution profiles such as that shown in Figure 13B were obtained. Using peak positions and variances obtained by fitting to a Gaussian Equation, the data listed in Table 3 were calculated. In these studies, more variability was observed when the Denizot-Delaage method was used for the calculation. As was observed for AVP dissociation, the rate constant for dimer dissociation obtained by analytical affinity chromatography was smaller than that observed for soluble dimers by several orders of magnitude. However, the relative rates of dissociation of AVP and BNP II from BNP II monomers were similar for the immobilized monomers and for monomers in solution (k_{NPII}:k_{AVP} = 1.8 and 1.1 respectively).

VI. CHARACTERIZATION OF THE MULTIPLE INTERACTIONS IN THE NEUROPHYSIN/NEUROHYPOPHYSIAL HORMONE SYSTEM — A CASE STUDY

A. Introduction

Analytical affinity chromatography provides a flexible general approach to correlate interactions in multimolecular systems including those for which different types of interactions modulate one another. There are no basic limitations on the size of molecular species which are immobilized or eluted as mobile interactants. Thus, a unified study can be made even when different interactions in the system involve molecules of sizes which obviate the use of other methods, such as equilibrium dialysis and ultracentrifugation. In addition, the chromatographic approach allows multiple interactions to be measured under similar experimental conditions (solvent, temperature, etc.), thus providing a unified quantitative description of a multimolecular system.

The study of multiple equilibria, which occur in the above-mentioned complexes between the neuroendocrine-peptide hormones oxytocin and vasopressin and the neurophysins (NP) and in the related complexes of the biosynthetic precursors of these peptides and proteins, represents a useful case study of the application of analytical-affinity chromatography.[43-48] Both peptide-protein and protein-protein interactions can be measured. Cooperative effects between these two types of interactions can be evaluated and the relationship between the cooperative properties of the hormone-NP complexes and the molecular organization of hormone/NP common biosynthetic precursors can be examined.

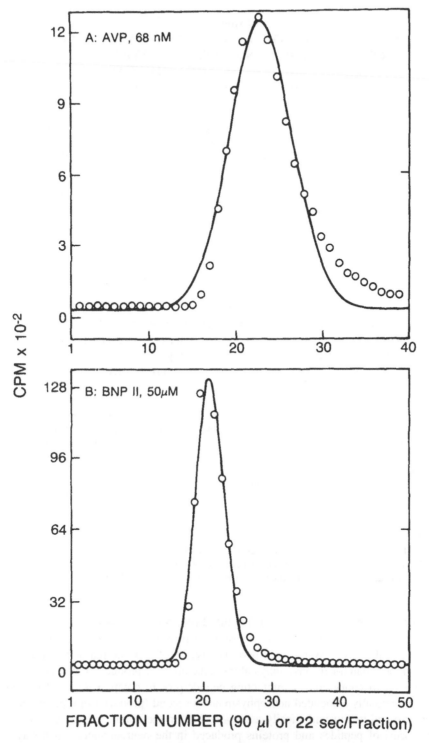

FIGURE 13. Zonal elutions, used for analysis of interaction kinetics, of [³H]Arg⁸-vaso-pressin and ¹²⁵I BNP II on affinity matrix of nonporous glass containing immobilized BNP II. The column (0.46 × 25 cm) was equilibrated and eluted at room temperature with 0.4 M ammonium acetate, pH 5.7, at a flow rate of 0.2 mℓ/min. A 200 μℓ sample was injected and 2-drop fractions (90 μℓ) were collected, starting at 0.5 min, directly in scintillation vials for counting. The solid lines represent computer-fits to Gaussian equations from which the peak variance was obtained. (A) Elution profile for 68 nM AVP. (B) Elution profile for 50 μM BNP II.

Table 2
CALCULATION OF THE KINETIC DISSOCIATION RATE CONSTANT FOR DISSOCIATION OF AVP FROM IMMOBILIZED BNP II USING THE HETHCOTE-DeLISI[a] OR THE DENIZOT-DELAAGE METHOD[b]

AVP conc. (nM)	V (ml)	V − V$_o$ (ml)	E(t′)[c] (sec)	Peak width[d] (sec)	W$_e$ (ml^2)	k$_d$ (sec^{-1}) H-D method	k$_d$ (sec^{-1}) D-D method
7.8	2.248	0.528	674.5	195.2	0.423	0.0083	0.0077
42.2	2.204	0.484	661.2	155.8	0.270	0.0120	0.0118
68	2.075	0.356	622.6	138.7	0.214	0.0111	0.0102

[a]   Equation 71.
[b]   Equation 68.
[c]   E(t$_o$) for an unretarded molecule was 548 ± 28 sec.
[d]   The peak width for an unretarded molecule was 60 ± 7 sec.

Table 3
CALCULATION OF THE KINETIC DISSOCIATION RATE CONSTANT FOR THE DISSOCIATION OF BNP II FROM IMMOBILIZED BNP II USING THE HETHCOTE-DeLISI[a] AND THE DENIZOT-DELAAGE METHODS[b]

BNP II conc. (μM)	V (ml)	V − V$_o$ (ml)	E(t′)[c] (sec)	Peak width[d] (sec)	W$_e$ (ml^2)	k$_d$ (sec^{-1}) H-D method	k$_d$ (sec^{-1}) D-D method
0.0375[e]	2.057	0.3249	617.2	77.0	0.0660	0.0328	0.100
0.0375[e]	1.947	0.2133	584.0	67.5	0.0507	0.0280	0.147
50	1.911	0.1773	573.2	77.4	0.0666	0.0177	0.024

[a]   Equation 71.
[b]   Equation 68.
[c]   E(t$_o$) for an unretarded molecule was 548 ± 28 sec.
[d]   The peak width for an unretarded molecule was 60 ± 7 sec.
[e]   Data are shown for two independent experiments carried out at [BNP II] = 0.0375 μM.

B. Biosynthetic Origin of Cooperative Peptide/Protein Complexes

Oxytocin, vasopressin, and the NPs are produced from common biosynthetic precursors both in the central nervous system and peripherally.[12-14,49-55] In the major pathway quantitatively, the hypothalamo-neurohypophysial tract, the ultimate products of biosynthesis (Figure 14) are a set of cooperatively interacting peptide/protein complexes, one for oxytocin and its biosynthetically associated neurophysin and a second (produced in separate neurons) for vasopressin and its biosynthetically-associated NP. In the mature complexes, which act as storage forms of peptides and proteins produced in the neuroendocrine pathway, self-association of NP and hormone-NP interaction mutually modulate one another in a manner which likely is related to the molecular organization of the precursors from which they form (see Figure 14). It has been possible to use the analytical-affinity chromatographic method to characterize the mature complexes and to begin to correlate the properties of these forms to molecular properties of the biosynthetic precursors.

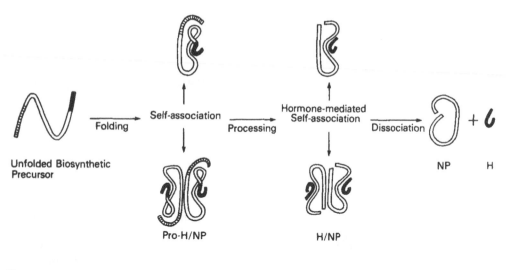

Unfolded Biosynthetic Precursor

Folding → Self-association → Processing → Hormone-mediated Self-association → Dissociation → NP + H

Pro-H/NP

H/NP

Biosynthesis-Packaging   Transport-Storage   Exocytosis

FIGURE 14. Scheme depicting relationship of biosynthetic precursor structure to molecular events occurring in neurohypophysial hormone (H)/neurophysin (NP) biosynthesis. The filled and open lines denote hormone and neurophysin segments, respectively. The hatched line represents the C-terminal glycopeptide occurring in pro-AVP/BNP II. Folding of the precursors is visualized to lead to self-association through the NP domains of the precursors. The NP-NP and H-NP interaction surfaces are retained after enzymatic processing, the latter of which leads to formation of noncovalent complexes between H and NP as well as dimers in secretory granules until released exocytotically. (From Kanmera, T. and Chaiken, I. M., *J. Biol. Chem.*, 260, 8474, 1985. With permission.)

C. Characterization of Mature Neuropeptide Hormone/Neurophysin Complexes

Two types of affinity matrices have been used in the examination of interactions in the neuropeptide hormone/NP system. As shown in Figure 15, elution of NP on immobilized-peptide ligand allows measurement of NP-ligand interactions. Soluble hormone-NP interactions as well as the effects of NP self-association on ligand binding can be evaluated. Both immobilized vasopressin (Lys 8-linked)[56] and immobilized Met-Tyr-Phe (α-carboxyl-linked, the tripeptide is a NP-binding analog of the N-terminal sequence of vasopressin)[43] have been available for these studies. Elutions on immobilized NP also can be performed (Figure 15) allowing measurement of self-association and the effects of peptide ligand binding on self-association.

Zonal elution of BNP II (vasopressin-associated) on Met-Tyr-Phe-aminoalkyl-agaroses shows behavior which fits a cooperative binding scheme distinct from strictly monovalent or bivalent models. Elution volumes do decrease generally as the concentration of competitive peptide (e.g., oxytocin or vasopressin) is increased in the elution buffer, as for example in Figure 16. However, when sets of competitive-elution data are plotted as $1/(V - V_o)$ vs. [L], curvilinearity at low [L] is observed in most cases (Figure 17). The reason for this behavior rests with the fact that NP exists as a mixture of low-affinity monomer and high-affinity dimers, with the degree of dimerization increased by peptide-ligand binding. Thus, while elution of NP at zero-soluble ligand reflects affinity-matrix binding of a mixture containing a significant amount of monomers, the more-retarded-than-expected elution in the presence of low concentrations of soluble peptide reflects elution of mixtures progressively more enriched in high-affinity dimers. Thus, elution volumes decrease more gradually, with increasing but low soluble-peptide concentrations, than expected from the degree of competition of the soluble peptides with affinity matrix for binding to the mobile NP.[44] Equilibrium-binding constants can be calculated from these data at both [L] = 0 and at [L] $\neq$ 0, as given in Table 4 (see also below).

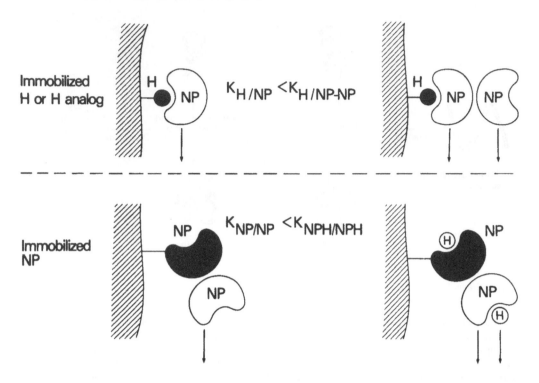

FIGURE 15.   Schematic diagram of quantitative affinity chromatographic analyses of interactions in the neuro-hypophysial hormone (H)/neurophysin (NP) system carried out with affinity matrices containing immobilized peptide hormone or hormone analog (H analog) and immobilized neurophysin. Affinity matrix preparations have been described.[43,44,56]

The higher peptide affinity of dimers than monomers also is expressed as a strong dependence of NP-elution volume on the amount (i.e., concentration) of NP in the initial zone. Thus, as shown in Figure 18, increasing zonal protein concentrations leads to increased retardation. Values of $1/K_{M/NP}$ vary sigmoidally with zonal concentration of NP (Figure 19), allowing an estimation of immobilized tripeptide ligand affinities for NP monomer (plateau at lowest concentrations of mobile NP) and dimer (plateau at high concentration of mobile NP).

NP dimerization has been measured by elution of NP on immobilized NP. As shown in the Figure 20 inset, zonal elution of [^{125}I]BNP II in buffer (containing no soluble-peptide ligand) shows a retardation displaced sufficiently from V_o to allow calculation of a dimerization constant as $K_{M/P}$ (Table 4). By including soluble-peptide ligand in the elution buffer, dimerization-dissociation constants also can be determined for partially- or fully-liganded NP. The main-elution profile of Figure 20 shows the case of close-to-saturating lysine vasopressin. For this liganded case, $K_{M/P}$ is much smaller (higher affinity) than that in buffer alone (Table 4). Thus, here as with the peptidyl-affinity matrix, zonal elution of NP can be used to measure the degree of modulation of binding affinities in the cooperative complexes between NPs and peptide hormones.

Analytical affinity chromatographic elutions on immobilized-peptide ligand and NP matrices allow calculation of a set of dissociation constants for self-association and peptide-ligand binding. These constants, compiled in Table 5, fit with the cooperativity linkage diagram, in Figure 21, which describes the interrelationship between noncovalent interactions which occur in mature neuropeptide hormone/NP complexes. The relationship of binding parameters argues that in the secretory granules in which these complexes are produced and stored the polypeptide system is driven to the fully self-associated, fully-liganded form.

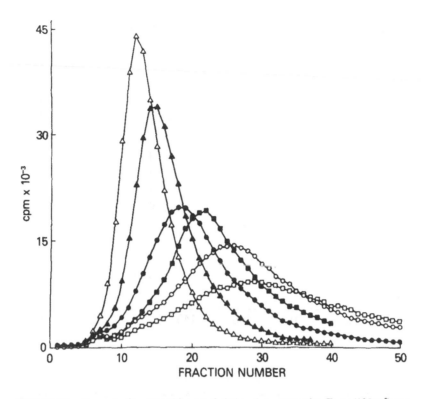

FIGURE 16. Neurophysin-competitive zonal elution chromatography. Zones (100 μℓ) containing ^{125}I-labeled NP II (340,000 cpm) and unlabled NP II (41.5 μg) were eluted on the affinity adsorbent Met-Tyr-Phe-AB-A (2.5-mℓ bed volume) in the presence of varying amounts of the soluble competing ligand LVP. Each continuous profile represents an experiment at the following LVP concentration (μM): (△) 64.2; (▲) 42.8; (●) 27.9; (■) 17.1; (○) 9.3: (□) 0. Other operating conditions were as follows: room temperature, gravity flow, 0.5 mℓ/fraction, and 0.4 M ammonium acetate buffer, pH 5.7. (From Angal, S. and Chaiken, I. M., *Biochemistry*, 21, 1574, 1982. With permission.)

The matrix systems used as described above to study the NP II-NP II self-association and NP II-peptide ligand interactions also can be used to evaluate binding characteristics of other native NPs (including formation of mixed hybrids), sequence-modified NPs, and peptide-hormone analogs. For example, the ability of different species of NPs to self-associate with one another can be evaluated directly by the degree of retardation on immobilized NP or by the degree to which the species induces increased retardation of a labeled NP on immobilized ligand. Figure 22 shows a direct demonstration that BNP I (oxytocin-associated in vivo) can form mixed hybrids with BNP II (vasopressin-associated in vivo), with a $K_{NP\ II/NP\ I}$ of $5.6 \times 10^{-6} M$ vs. $K_{NP\ II/NP\ II}$ of $1.3 \times 10^{-5} M$ for NP II homologous dimerization (Figure 23). These and related results have shown that the self-association surface is a conserved structural feature among native NPs of different sequence and species. In addition, affinity-chromatographic behavior of analogs on both immobilized NP and immobilized-peptide ligand have been used to show that active-site photolabeled NP II has a greatly reduced ability to self-associate with immobilized NP II (see Figure 23 and 24). The results with photolabeled NP have led to the conclusion that the self-association surface is distorted in the derivative (when the ligand-binding site is covalently occupied) and also to the general view that cooperativity between the noncovalent peptide-ligand binding and self-association surfaces can be used as a sensitive signal of molecular organization in native NP.

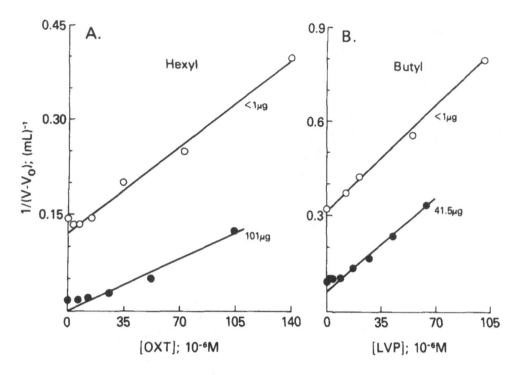

FIGURE 17.   Linearized plots of competitive zonal-elution data for ^{125}I-labeled bovine neurophysin II ([^{125}I]BNP II) on Met-Tyr-Phe-aminohexyl-(A) and -aminobutyl-(B) agarose (columns were 2-mℓ bed volumes packed in 7 mm internal diameter columns). Elution volumes (V) for the elution series shown in Figure 3 and three other similar series (not shown) are plotted as $1/(V - V_o)$ vs. the concentrations (as indicated in the figure) of competitive ligand ([L]) which in this case is either of the neuropeptide hormones, oxytocin (OXT), or lysine-vasopressin (LVP). Open circles are for elutions at low-total protein in zones applied to the column (<1 μg of [^{125}I]BNP II only); closed circles are for high total amounts of protein applied to the column 1 μg of [^{125}I]BNP II plus either 101 μg or 41.5 μg of unlabeled BNP II). Straight lines were drawn according to linear least-squares regression analysis of data points except for points at [L] = 0 and for values of [L] near 0 which deviate from linearity. Dissociation constants, $K_{P/L}$ and $K_{M/P}$, calculated from the linearized plots using Equations 11 and 12 are shown in Table 4. (From Angal, S. and Chaiken, I. M., *Biochemistry*, 21, 1574, 1982. With permission.)

D. Precursor Interactions and Molecular Organization

The quantitative differences in self-association of liganded vs. unliganded NP allow predictions about the chromatographic behavior of biosynthetic precursors that can be used to gain insight into the molecular organization of these latter molecules. If the precursors fold into well-defined conformations it is likely that these would mimic the structure of liganded NP. This seems especially likely with the oxytocin/NP precursor which is composed mainly of hormone and NP domains with a tripeptide-spacer linkage between them and a single His residue at the carboxyl terminus (Figure 25). Thus, if folded precursors have accessible self-association surfaces, the precursor would be expected to self-associate and the affinity of this self-association would be close to that of liganded NP.

This idea was tested through analytical affinity chromatography using [di-acetimidyl 30,71, des His 106]pro-OT/NPI, a semisynthetic oxytocin/BNP I precursor analog which differs from the native precursor only in missing C-terminal His 106 (assumed unlikely to be important for self-association) and having acetimidyl-protecting groups at the two ε-NH$_2$s in the NP domain (see Figure 25). The data of Figure 26A and B show that di-acetimidyl-NP I can associate with immobilized NP II, with affinities in the presence and absence of hormone similar to the corresponding values for unprotected NP II. The chromatographic retardation data for semisynthetic pro-OT/NP I (Figure 26A and B) show that the precursor

Table 4

REPRESENTATIVE ZONAL-ELUTION ANALYTICAL AFFINITY CHROMATOGRAPHIC DATA OBTAINED FOR NEUROPHYSIN SELF-ASSOCIATION AND NEUROPHYSIN-PEPTIDE INTERACTIONS USING IMMOBILIZED PEPTIDE AND NEUROPHYSIN-AFFINITY MATRICES

Mobile interactant	Amount in zone (μg)	Ligand immobilized on affinity matrix	$[M]_T$	Competitive ligand or effector (concentration)	Equation 12 K_{MP} (M) calculated at $[L] = 0$ (with or without effector)	Equation 11 K_{MP} (M) calculated at $[L] \neq 0$	$K_{P/L}$ (M)	Ref.
[125I]-BNP II	<1	Met-Thr-Phe-aminobutyl-	3.0×10^{-4}	LVP (varying)	4.8×10^{-5}	4.7×10^{-5}	6.6×10^{-5}	44
	<1	Met-Tyr-Phe-aminohexyl-	5.9×10^{-4}	OT (varying)	5.1×10^{-5}	4.3×10^{-5}	6.2×10^{-5}	44
	41.5	Met-Tyr-Phe-aminobutyl-	3.0×10^{-4}	LVP (varying)	1.4×10^{-5}	8.9×10^{-6}	1.4×10^{-5}	44
	101	Met-Tyr-Phe-aminohexyl-	5.9×10^{-4}	OT (varying)	6.1×10^{-6}	7.0×10^{-7}	1.8×10^{-6}	44
[125I]-BNP II	<1	BNP II-	1×10^{-4}	None	1.4×10^{-5}			45
	<1	BNP II-	1×10^{-4}	None	1.6×10^{-5}			44
				LVP ($1 \times 10^{-4} M$)	4.4×10^{-7}			45
[14C-diAcet]BNP I	<1	BNP II-	7×10^{-5}	None	7.7×10^{-6}			48
				LVP ($1 \times 10^{-4} M$)	5.9×10^{-8}			48
[14C-diAcet, desHis 106]-pro-OT/BNP I	<1	Met-Tyr-Phe-	8.5×10^{-3}	None	1.2×10^{-4}			48
	<1	Met-Tyr-Phe-	8.5×10^{-3}	None	5.3×10^{-3}			48
	<1	BNP II-	7×10^{-5}	None	5.9×10^{-7}			48
				LVP ($1 \times 10^{-4} M$)	7.1×10^{-8}			48
[125I]-BNP II	2.5	BNP II-	1×10^{-4}	None	1.3×10^{-5}			46
[125I]-BNP I	1	BNP II-	1×10^{-4}	None	5.6×10^{-6}			46

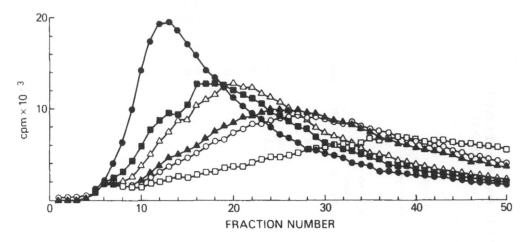

FIGURE 18. Effect of concentration of applied bovine neurophysin II (BNP II) on zonal-elution behavior on Met-Tyr-Phe-aminobutyl-agarose. Zones (100 μℓ) containing <1 μg of [^{125}I]BNP II and unlabeled BNP II (varying amounts as specified) were eluted in 0.4 M ammonium acetate, pH 5.7, in the absence of soluble-competitive ligand. Each continuous profile represents a separate elution with the following amounts (in micrograms) of added unlabeled BNP II per zone: 0 (●), 6.35 (■), 10.3 (△); 20.7 (▲), 41.3 (○), 82.6 (□). Elutions were carried out at room temperature. (From Angal, S. and Chaiken, I. M., *Biochemistry*, 21, 1574, 1982. With permission.)

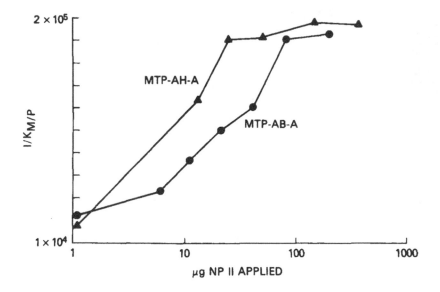

FIGURE 19. Affinity of neurophysin for immobilized ligands as a function of protein concentration. Elution volumes (V) obtained with Met-Tyr-Phe-AB-A (from Figure 3) and with Met-Tyr-Phe-AH-A (not shown) were substituted into Equation 1 to yield association constants 1/K_{MP} of NP II for the immobilized ligands. Details are as described in the legend to Figure 18. (From Angal, S. and Chaiken, I. M., *Biochemistry*, 21, 1574, 1982. With permission.)

can bind to immobilized NP, with the latter in either the unliganded (A) or liganded (B) form. Most notable is that the quantitative degree of association of precursor to unliganded immobilized NP II is greater than that of unliganded NP II to unliganded-immobilized NP II (Figure 26A) and the association of precursor to liganded-immobilized NP II is closely related to that of liganded NP II to liganded-immobilized NP II (Figure 26B). The enhanced self-association potential of neuropeptide hormone/NP precursor was also observed for pulse-

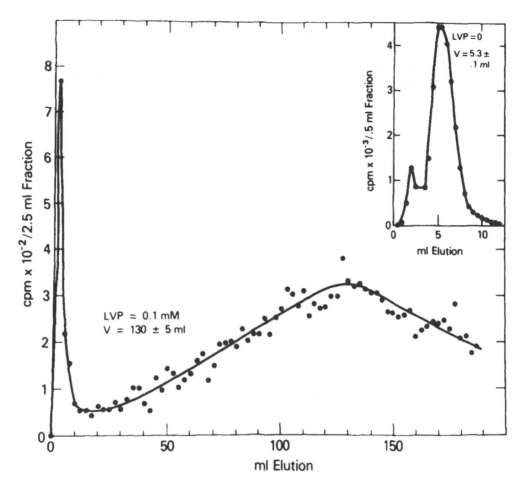

FIGURE 20. *Elution behavior of* [^{125}I]BNP II *on* [BNP II]-Sepharose *in the presence and absence of soluble peptide ligand, lysine-vasopressin (LVP). Zones (300 µℓ) containing ~ 1 µg of* [^{125}I]BNP II *were applied to* [BNP II]-Sepharose *(1.5 mℓ bed volume,* [M]$_T$ = 1.4 × 10^{-4} M*) equilibrated with 0.4 M ammonium acetate, pH 5.7, containing 0.1 mM LVP (main diagram) or in the absence of added LVP (inset). The zonal-elution volumes were used to calculate* K$_{M/P}$ *(dissociation constants for protein-protein interaction) using Equation 12. These values are shown in Table 4. (From Chaiken, I. M., Tamaoki, H., Brownstein, M. J., and Gainer, H.,* FEBS Lett., *164, 361, 1983. With permission.)*

labeled rat provasopressin/NP by showing binding of the latter to unliganded immobilized NP II.[44] The affinity chromatographic results argue that the neuropeptide hormone/NP precursors, with self-association properties similar to liganded NP, fold into structures with a well-defined intramolecular-hormone domain-NP domain contact mimicking the intermolecular hormone-NP interaction in mature complexes (Figure 14). In secretory granules the biosynthetic precursors for oxytocin, vasopressin, and neurophysins likely form self-associated complexes and it is in these self-associated forms that the precursors likely are processed enzymatically to produce mature-neuropeptide hormones and NPs. Such a view helps define the molecular transitions which occur in the class of peptide-secreting cells for neurohypophysial hormones and NPs (Figure 27).

The folding properties of neuropeptide/NP precursors also can be established by analytical affinity chromatography. The proposed mechanism of biosynthetic origin (Figure 14) predicts that the precursors, with eight potential disulfide bonds, fold spontaneously upon in vivo translation to form the native disulfide-bonded form. Given this prediction, prohormone/NP, like proinsulin, chymotrypsinogen, and other disulfide-containing precursors, should

Table 5

**SUMMARY OF DISSOCIATION CONSTANTS FOR
NEUROPHYSIN/NEUROPEPTIDE HORMONE COMPLEXES AND
BIOSYNTHETIC PRECURSORS DETERMINED BY
ANALYTICAL-AFFINITY CHROMATOGRAPHY**

Dissociation constant	Interaction process (immobilized component/mobile component)	Value of $K_{M/P}$ (M)	Ref.
$K_{P/P}$	BNP II/[^{125}I]BNP II	1.4×10^{-5}	45
		1.6×10^{-5}	44
		1.3×10^{-5}	46
	BNP II/[^{125}I]BNP I	5.6×10^{-6}	46
	BNP II/[^{14}CdiAcet]BNP I	7.7×10^{-6}	48
$K_{PL/PL}$	BNP II/[^{125}I]BNP II + LVP	4.4×10^{-7}	45
	BNP II/[^{14}CdiAcet]BNP I + LVP	5.9×10^{-8}	48
$K_{P/Pro}$	BNP II/[^{14}CdiAcet]Pro-OT/BNP I	5.9×10^{-7}	48
$K_{PL/Pro}$	BNP II/[^{14}CdiAcet]Pro-OT/BNP I + LVP	7.1×10^{-8}	48
$K_{P/L}$	Met-Tyr-Phe/[^{125}I]BNP II (low µg in Figure 19; no competitor)	$0.5 - 1 \times 10^{-4}$	44
	Met-Tyr-Phe/[^{125}I]BNP I (no competitor)	1.2×10^{-4}	46
$K_{PP/L}$	Met-Tyr-Phe/[^{125}I]BNP II (high µg in Figure 9A; no competitor)	1×10^{-5}	44
$K_{PPL/L}$	Met-Tyr-Phe/[^{125}I]BNP II (high µg; with competitor)	7×10^{-7}	44

be able to maintain correct disulfides in disulfide-interchange conditions. NP (with seven disulfides), like insulin and chymotrypsin, cannot maintain native disulfides under interchange conditions, indicating an inability to fold spontaneously.[48,57,58] Chromatographic analysis of semisynthetic pro-OT/NP I on immobilized NP II after treatment with reducing agent shows no decrease in self-association (Figure 28B and C). In contrast, NP so treated loses it ability to bind peptide ligand (Figure 28A). The results argue that, as shown in Figure 14, hormone/NP-precursor folding ensues upon biosynthetic assembly, providing a form which self-associates upon packaging into secretory granules before enzymatic processing within granules.

VII. SUMMARY OF EQUATIONS FOR ANALYSIS OF MOLECULAR INTERACTIONS BY ANALYTICAL-AFFINITY CHROMATOGRAPHY

This section summarizes the relationships which can be used to analyze experimental data obtained for various types of analytical-affinity chromatographic systems. The equation numbers are as defined when cited originally in the chapter.

A. Competitive-Elution Analysis of Monovalent and Bivalent Interactions

In this type of experiment, the mobile interactant P is eluted on a matrix with immobilized L, defined as M, in the absence or presence of soluble L. The elution volume of P is determined as a function of the concentration of L in the mobile phase.

1. Zonal elution chromatography of a monovalent interactant

$$\frac{V_o - V_m}{V - V_o} = \frac{K_{M/P}}{[M]_T} + \frac{K_{M/P}}{K_{L/P}} \frac{[L]_T}{[M]_T} \tag{11}$$

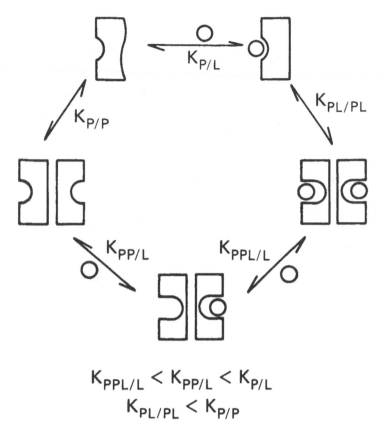

$$K_{PPL/L} < K_{PP/L} < K_{P/L}$$
$$K_{PL/PL} < K_{P/P}$$

FIGURE 21. Scheme of cooperative relationship between peptide ligand (○) binding and protein self-association in the neurophysin/hormone system. $K_{P/L}$, $K_{PP/L}$, and $K_{PPL/L}$ are affinity constants of ligand for neurophysin monomer, unliganded dimer, and singly liganded dimer, respectively. $K_{P/P}$ and $K_{PL/PL}$ are dissociation constants for self-association of, respectively, unliganded and liganded neurophysin monomers. The scheme denotes the relationship between intermolecular hormone binding and self-association occurring in mature hormone-neurophysin noncovalent complexes. A similar relationship pertains for precursor between intramolecular hormone domain-neurophysin domain interaction and precursor self-association. (Adapted from Chaiken, I. M., Abercrombie, D. M., Kanmera, T., and Sequeira, R. P., in *Peptide and Protein Review*, Vol. 1, Hearn, M. T. W., Ed., Marcel Dekker, New York, 1983, 139. With permission.)

2. Zonal elution chromatography of a bivalent interactant

$$\frac{V_o - V_m}{V - V_o} = \frac{1 + 2\left(\frac{[L]}{K_{P_2/L}}\right) + \left(\frac{[L]}{K_{P_2/L}}\right)^2}{2\left(\frac{[M]_T}{K_{M/P_2}}\right) + \left(\frac{[M]_T}{K_{M/P_2}}\right)^2 + 2\left(\frac{[L][M]_T}{K_{P_2/L} K_{M/P_2}}\right)} \quad (15)$$

3. Continuous elution chromatography of a monovalent interactant

$$\bar{V}_i = V_o + K_{L/P} \frac{\bar{V}_{lim} - \bar{V}_i}{[L]_T} \quad (59)$$

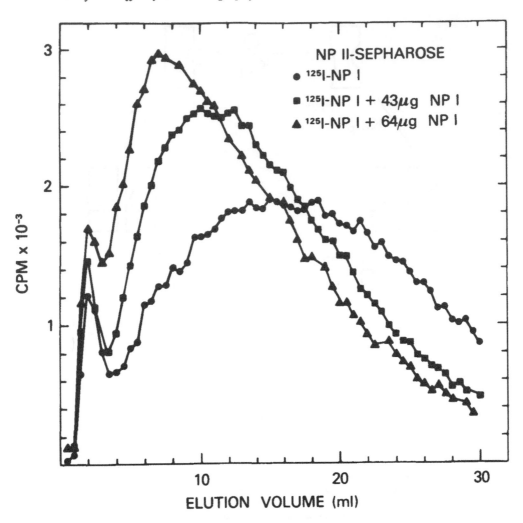

FIGURE 22. Detection of mixed hybrid formation between BNP I and II. Zones (200 $\mu\ell$) containing [^{125}I]BNP I (~1 μg, 10^5 cpm) plus the indicated amounts of added unlabeled BNP I were eluted from [BNP II]-Sepharose with 0.4 *M* ammonium acetate, pH 5.7. Reduced V of [^{125}I]BNP I with increasing unlabeled BNP I connotes competitiveness between mixed and homologous dimer formation.

B. Characterization of Self-Association (Dimerization) and the Effect of Ligand Binding

Immobilization of a protein which exhibits self-association permits a characterization both of that association and the effects on the equilibrium from binding additional ligands. Using a column of immobilized protein monomer, the elution volume of soluble protein is determined as a function of its concentration either in the absence of an effector ligand or in the presence of saturating amounts of the effector.

1. Zonal-elution chromatography of an associating protein in the absence of effector ligands

$$\frac{V_o - V_m}{V - V_o} = \frac{K_{M/P}}{2[M]_T} + \left[\frac{K_{M/P}}{2}\left(1 + \frac{8[P]_T}{K_{P/P}}\right)^{1/2} + [P]_T\right]\frac{1}{[M]_T} \qquad (28)$$

$$\frac{V_o - V_m}{V - V_o} = \frac{K_{M/P}}{[M]_T} \quad (\text{when } [P]_T \ll K_{P/P} \text{ and } K_{M/P}) \qquad (29)$$

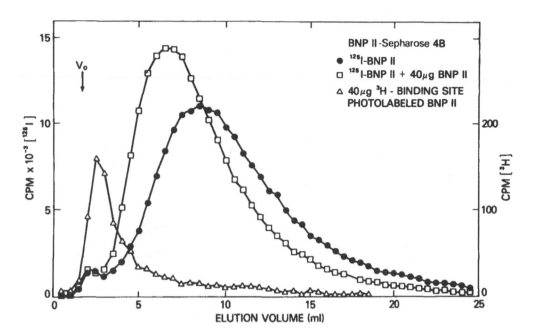

FIGURE 23. Affinity chromatography of native and binding site photolabeled BNP II on [BNP II]-Sepharose. Zones (250 μℓ) containing 40 μg of [³H]-labeled binding-site photolabeled BNP II (3 × 10³ cpm) alone (△) and [¹²⁵I]BNP II (2.5 μg specific activity ~0.8 × 10⁵ cpm/μg) alone (●) with 40 μg of added native BNP II (□) were eluted from the immobilized BNP II with 0.4 *M* ammonium acetate, pH 5.7. V_o = 1.7 mℓ. (From Abercrombie, D. M., Kanmera, T., Angal, S., Tamaoki, H., and Chaiken, I. M., *Int. J. Pept. Prot. Res.*, 24, 218, 1984. With permission.)

2. Zonal-elution chromatography of an associating protein in the presence of saturating concentrations of effector ligand

$$\frac{V_o - V_m}{V - V_o} = \frac{K_{ML/PL}}{2\,[M]_T} + \left[\frac{K_{ML/PL}}{2}\left(1 + \frac{8[P]_T}{K_{PL/PL}}\right)^{1/2} + [P]_T\right]\frac{1}{[M]_T} \tag{33}$$

$$= \frac{K_{ML/PL}}{[M]_T} \quad \text{when} \quad [P]_T << K_{ML/PL} \text{ and } K_{PL/PL} \tag{34}$$

C. Characterization of Binding of a Soluble Small-Molecule Interactant with an Immobilized Macromolecule

In this type of chromatography, the interaction between a soluble interactant and the immobilized molecule is characterized from the retardation of its elution.

1. Zonal elution chromatography of a ligand which binds to a monovalent immobilized molecule

$$\frac{V_o - V_m}{V - V_o} = \frac{K_{M/L}}{[M]_T} + \frac{[L]_T}{[M]_T} \tag{38}$$

2. Continuous elution chromatography of a ligand which binds to a monovalent immobilized molecule

$$\frac{1}{\overline{V} - V_o} = \frac{K_{M/P}}{M_T} + \frac{[P]^o}{M_T} \tag{61}$$

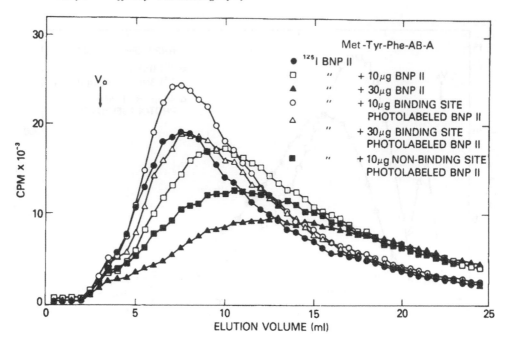

FIGURE 24. Affinity chromatographic analysis of dimerization-derived enhancement of [^{125}I]BNP II binding to Met-Tyr-Phe-AB-agarose by native BNP II, binding-site photolabeled BNP II, and nonbinding-site photolabeled BNP II (a mixture of active, photolabeled BNP II and native BNP II). Zones (220 $\mu\ell$) containing [^{125}I]BNP II (~1 μg, specific activity ~ 0.5 $\times$ 10^6 cpm/μg) alone or with the added proteins indicated were eluted from the Met-Tyr-Phe-AB-agarose with 0.4 M ammonium acetate, pH 5.7. V$_o$ = 3.0 mℓ. (From Abercrombie, D. M., Kanmera, T., Angal, S., Tamaoki, H., and Chaiken, I. M., *Int. J. Pept. Prot. Res.*, 24, 218, 1984. With permission.)

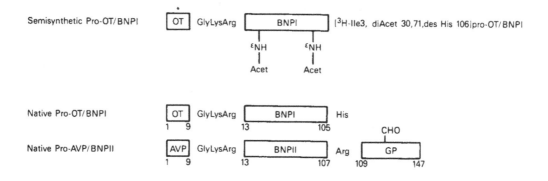

FIGURE 25. Schematic diagram of linear sequences of active bovine neurohypophysial hormone/neurophysin biosynthetic precursors (pro-OT/BNP I and pro-AVP/BNP II) and a semisynthetically-derived pro-OT/BNP I analog.[47] The native precursor schemes are based on cDNA sequence data[49] and emphasize the arrangement of sequence domains, NP, AVP, OT, and GP (glycopeptide), each of which accumulates as a mature form in secretory granules after posttranslational processing. The native precursors also contain intervening and terminal amino acid residues, GKR (Gly-Lys-Arg), R (arginine), and H (histidine). The semisynthetic precursor has been produced by chemical stitching of oxytocinyl-Gly-Lys-Arg and BNP I domains.[48] * depicts tritium in Ile. The numbers in parentheses indicate precursor residues starting at hormone residue 1/2 Cys 1. (Adapted from Kanmera, T. and Chaiken, I. M., *J. Biol. Chem.*, 260, 8474, 1985. With permission.)

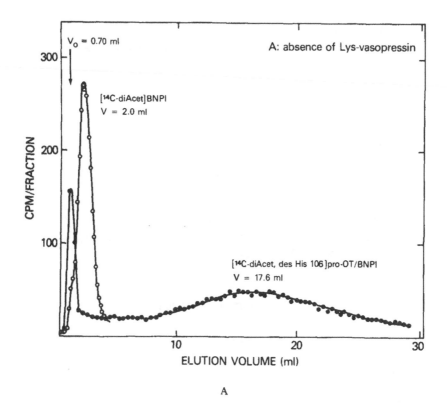

A

FIGURE 26. Analytical affinity chromatography elutions of semisynthetic precursor on [BNP II]Sepharose. Zones containing 1500 to 3000 cpm (<0.5 µg) of [^{14}C-diAcet]BNP I and [^{14}C-diAcet, des His 106]pro-OT/BNP I were eluted on [BNP II]Sepharose (70 nmol BNP II/mℓ of bed volume, 198 µℓ bed volume; prepared before[11]) with 0.4 M ammonium acetate containing 0.5% BSA (pH 5.7) in the absence (A) and the presence (B) of 0.1 mM LVP. Other conditions were as follows: Fraction size: (A) 4 drops (157 µℓ) for BNP I and 10 drops (391 µℓ) for the precursor (B) 60 drops (2.35 mℓ) for both elutions. Flow rate: 5 mℓ/hr; Temperature: ambient. (From Kanmera, T. and Chaiken, I. M., *J. Biol. Chem.*, 260, 8474, 1985. With permission.)

3. Zonal elution chromatography of a ligand which binds to a multivalent immobilized molecule

$$\frac{V - V_o}{V_o - V_m} = \frac{[M]_T \, \bar{\nu}}{[L]} \approx \frac{[M]_T}{K_1} \quad \text{(at low [L])} \tag{51,52}$$

VIII. GLOSSARY OF TERMS

A. For Zonal Analytical Affinity Chromatography

L   Soluble interacting molecule
P   Mobile interactant
M   Immobilized interactant
$K_{I,J}$ Dissociation constant of binary complex of molecules I and J
σ_i Equilibrium partition coefficient for molecule interacting with affinity matrix
$\sigma_{o,i}$ Equilibrium-partition coefficient for a molecule unretarded on affinity matrix
α Denotation of stationary phase

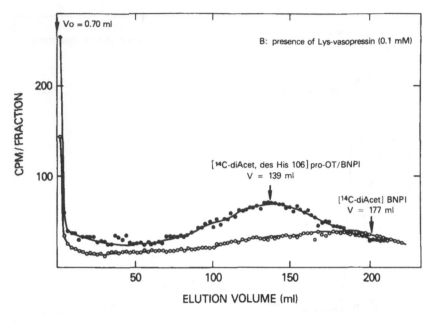

FIGURE 26B

β   Denotation of mobile phase
V   Elution volume of mobile interactant
V_o Elution volume of unretarded molecule
V_m Volume of the mobile phase
V_S Volume of the stationary phase (pore volume in the case of porous matrix)
[M]   Concentration of unliganded-immobilized interactant
$[M]_T$ Total concentration of immobilized interactant
$[P]_T$ Total concentration of mobile interactant
$[P]_T^\circ$ Total initial concentration of a mobile interactant (in zonal elution, the initial concentration in the zone introduced to top of affinity column)
[L]   Concentration of soluble, free-interacting molecule
$K_{IJ^*/K}$ Dissociation constant of a ternary complex IJK into IJ + K in which dissociation of molecule K is from molecule J in the complex
$\bar{\nu}$ Moles of L bound per mole of M
k_i Intrinsic-association constant for the "ith" class of sites for a molecule exhibiting multiple-binding equilibria

B. For Continuous-Elution Analytical Affinity Chromatography

$\bar{V}$ First moment of the advancing boundary of mobile interactant in continuous-elution chromatography (hypothetically the boundary at which the concentration of mobile component $[P] = [P]^\circ/2$)
$\bar{V}_{lim}$ Limiting value for the first moment V as the concentration of mobile interactant approaches zero
k_{IJ} Association constant for formation of a binary complex IJ from molecules I and J
f   Valency of multivalent-mobile interactant
K_{AV} Partition coefficient of mobile component P with affinity matrix

C. For Kinetics Evaluation of Zonal-Analytical Affinity Chromatography
k_a Kinetic association rate constant

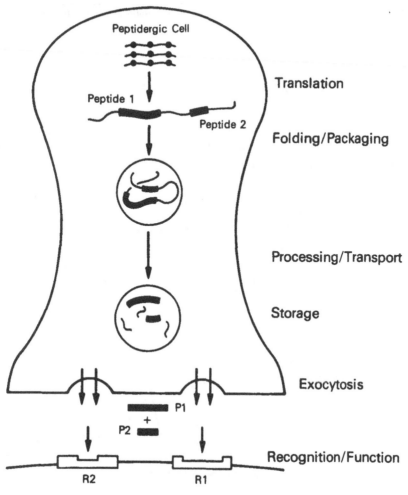

FIGURE 27. Schematic diagram depicting the molecular events occurring in neurophysin/ neurohypophysial hormone (NP/H) pathways. Overall processes of biosynthesis, packaging into neurosecretory granules, and proteolytic processing of neurophysin/hormone precursors are shown. Precursors are viewed as folded to form intramolecular interactions between hormone and neurophysin domains, with these molecules able to self-associate. The latter interaction may modulate the concentrations of hormones within granules as well as precursor processing events. Proteolytic processing within granules leads to mature hormone-neurophysin noncovalent complexes which also self-associate to an extent which is enhanced by hormone binding. The molecular events viewed to occur in such pathways are based largely on characterization of the most abundant case, hypothalamo-neurohypophysial neurons. The degree of similarity of this scheme to molecular mechanisms in other hormone/neurophysin pathways is not yet fully established. (From Chaiken, I. M., Kanmera, T., and Sequeira, R. P., in *Opioid Peptides: Molecular Pharmacology, Biosynthesis, and Analysis*, Rapaka, R. S. and Hawks, R. L., Eds., NIDA Press, 1986, 3. With permission.)

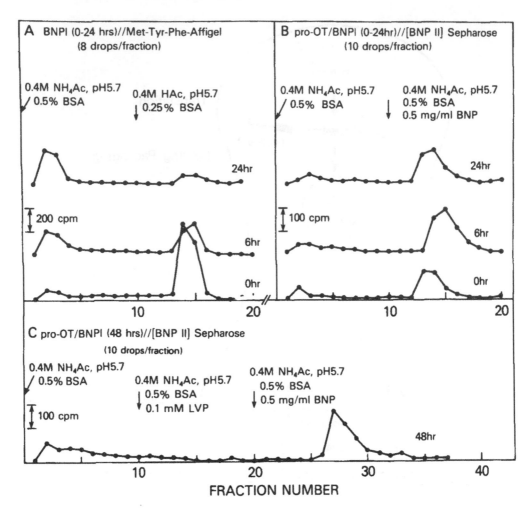

FIGURE 28. Analytical affinity chromatography analyses of sulfhydryl-treated BNP I and semisynthetic precursor. (A) Elutions of [¹⁴C-diAcet]BNP I, treated for 0, 6, and 24 hr with 16 μ*M* dithiothreitol (DTT), on Met-Tyr-Phe-AffiGel-102 (0.3 × 4.4 cm, 8.5 μmol tripeptide/m*ℓ* settled gel). Fractions of 10 drops each were collected. (B) Elution of [¹⁴C-diAcet, des His 106]pro-OT/BNP I, treated for 0, 6, and 24 hr with 16 μ*M* DTT (see Experimental) and then gel filtered to remove DTT, on [BNP II]Sepharose (0.3 × 1.8 cm, 72 μmol BNP II/m*ℓ* settled gel). Fractions of 8 drops each were collected. (C) As in B, except that twice the volume of precursor reaction mixture was analyzed and an intermediate elution step with LVP was added as shown. (From Kanmera, T. and Chaiken, I. M., *J. Biol. Chem.*, 260, 8474, 1985. With permission.)

k_d Kinetic dissociation rate constant

$E(t_o)$ Peak position (in time units) for unretarded mobile molecule

$E(t')$ Peak position (in time units) for mobile interactant

σ_o Standard deviation of peak of unretarded mobile molecule

σ' Standard deviation of peak of mobile interactant

k_{-1} Mass transfer desorption rate constant

W_e Peak variance

F   Volumetric flow rate

T   Sample thickness

h   Column length

IX. CONCLUDING COMMENTS

Analytical affinity chromatography has been found to provide an effective methodology to measure quantitative properties of macromolecular interactions. A close quantitative relatedness of chromatographically obtained equilibrium and analogous constants determined fully in solution has been found for a growing number of proteins. This consistently observed correlation has formed the basis for expanding theoretical treatments which describe the chromatographic behavior not only of monovalent-molecular systems of varying types but also of multivalently interacting macromolecules, including those which exhibit cooperativity. Quantitative use of affinity chromatography as a micromethod offers a biochemical-analysis tool to characterize biologically interesting molecules available in only relatively small amounts, a circumstance likely to become increasingly important given the rapidly expanding discovery of such species. The potential to measure chemical-rate constants by affinity chromatography has been recognized and experimental tests of the available theory are being made. While the relatedness of rate constants, derived chromatographically, to the molecular events of macromolecules in their biologically meaningful environments needs more study, the potential value of being able to measure such constants through analytical affinity chromatography makes further evaluation important. The development of affinity chromatographic systems, including designing matrices and elution conditions, historically has been dependent on having at least a rudimentary understanding of the interaction characteristics (most often in solution) of the isolated molecules to be utilized as the affinity matrix-mobile interactant system. The development of analytical affinity chromatography now offers the opportunity to use matrix-mobile interactant systems to study mechanisms of biomolecular interactions and therein to obtain an understanding of such interactions which often is not easily obtained by solution approaches alone.

REFERENCES

1. **Dunn, B. M. and Chaiken, I. M.,** Quantitative affinity chromatography. Determination of binding constants by elution with competitive inhibitors, *Proc. Natl. Acad. Sci. U.S.A.,* 71, 2382, 1974.
2. **Nichol, L. W., Ogston, A. G., Winzor, D. J., and Sawyer, W. H.,** Evaluation of equilibrium constants by affinity chromatography, *Biochem. J.,* 143, 435, 1974.
3. **Dunn, B. M. and Chaiken, I. M.,** Evaluation of quantitative affinity chromatography by comparison with kinetic and equilibrium dialysis methods for the analysis of nucleotide binding to staphylococcal nuclease, *Biochemistry,* 14, 2343, 1975.
4. **Kasai, K. and Ishii, S.,** Quantitative analysis of affinity chromatography of trypsin. A new technique for investigation of protein-ligand interactions, *J. Biochem.,* 77, 261, 1975.
5. **Chaiken, I. M.,** Quantitative uses of affinity chromatography, *Anal. Biochem.,* 97, 1, 1979.
6. **Chiancone, E., Gattoni, M., and Antonini, E.,** Subunit exchange chromatography: principles and applications, in *Affinity Chromatography and Biological Recognition,* Chaiken, I. M., Parikh, I., and Wilchek, M., Eds., Academic Press, Orlando, Fla., 1983, 103.
7. **Abercrombie, D. M. and Chaiken, I. M.,** Zonal elution quantitative affinity chromatography and analysis of molecular interactions, in *Affinity Chromatography — A Practical Approach,* Dean, P. D. G., Johnson, W. S., and Middle, F. A., Eds., IRL Press, Oxford, 1984, 169.
8. **Winzor, D. J.,** Quantitative characterisation of interactions by affinity chromatography, in *Affinity Chromatography — A Practical Approach,* Dean, P. D. G., Johnson, W. S., and Middle, F. A., Eds., IRL Press, Oxford, 1984, 149.
9. **Eilat, D. and Chaiken, I. M.,** Expression of multivalency in the affinity chromatography of antibodies, *Biochemistry,* 18, 790, 1979.
10. **Chaiken, I. M., Eilat, D., and McCormick, W. M.,** Appendix: derivation and evaluation of equations for independent bivalent interacting systems in quantitative affinity chromatography, *Biochemistry,* 18, 794, 1979.

11. **Nicolas, P., Batelier, G., Rholam, M., and Cohen, P.**, Bovine neurophysin dimerization and neuro-hypophysial hormone binding, *Biochemistry*, 19, 3565, 1979.

12. **Cohen, P., Nicolas, P., and Camier, M.**, Biochemical aspects of neurosecretion: neurophysin-neurohy-pophysial hormone complexes, *Curr. Top. Cell. Reg.*, 15, 263, 1979.

13. **Breslow, E.**, Neurophysin: biology and chemistry of its interactions, in *Cell Biology of the Secretory Process*, Cantin, M., Ed., S. Karger, Basel, 1984.

14. **Chaiken, I. M., Abercrombie, D. M., Kanmera, T., and Sequeira, R. P.**, Neuronal peptide-protein complexes: neurophysins and associated neuropeptide complexes, in *Peptide and Protein Review*, Vol. 1, Hearn, M. T. W., Ed., Marcel Dekker, New York, 1983, 139.

15. **Lagercrantz, C., Larsson, T., and Karlsson, H.**, Binding of some fatty acids and drugs to immobilized bovine serum albumin studied by column affinity chromatography, *Anal. Biochem.*, 99, 352, 1979.

16. **Tanford, C.**, *Physical Chemistry of Macromolecules*, John Wiley & Sons, New York, 1961, chap. 8.

17. **Chaiken, I. M. and Taylor, H. C.**, Analysis of ribonuclease-nucleotide interactions by quantitative affinity chromatography, *J. Biol. Chem.*, 251, 2044, 1976.

18. **Brodelius, P. and Mosbach, K.**, Determination of dissociation constants for binary dehydrogenase-coenzyme complexes by (bio)affinity chromatography on an immobilized AMP-analogue, *Anal. Biochem.*, 72, 629, 1976.

19. **Brinkworth, R. I., Masters, C. J., and Winzor, D. J.**, Evaluation of equilibrium constants for the interaction of lactate dehydrogenase isoenzymes with reduced nicotinamide-adenine dinucleotide by affinity chromatography, *Biochem. J.*, 151, 631, 1975.

20. **Nishikata, M., Kasai, K., and Ishii, S.**, Affinity chromatography of trypsin and related enzymes. IV. Quantitative comparison of affinity adsorbents containing various arginine peptides, *J. Biochem.*, 82, 1475, 1977.

21. **Dunn, B. M., Danner-Rabovsky, J., and Cambias, J. S.**, Application of quantitative affinity chromatography to the study of protein-ligand interactions, in *Affinity Chromatography and Biological Recognition*, Chaiken, I. M., Parikh, I., and Wilchek, M., Eds., Academic Press, Orlando, Fla., 1983, 93.

22. **Veronese, F. M., Bevilacqua, R., and Chaiken, I. M.**, Drug-protein interactions: evaluation of the binding of antipsychotic drugs to glutamate dehydrogenase by quantitative affinity chromatography, *Mol. Pharmacol.*, 15, 313, 1979.

23. **Liu, Y. C., Ledger, R., Bryant, C., and Stellwagen, E.**, Quantitative analysis of immobilized dye-protein interactions, in *Affinity Chromatography and Biological Recognition*, Chaiken, I. M., Parikh, I., and Wilcheck, M., Eds., Academic Press, Orlando, Fla., 1983, 135.

24. **Yon, R. J. and Kyprianou, P.**, Biospecific desorption from low-specificity adsorbents with emphasis on 10-Carboxydecylamino-Sepharose, in *Affinity Chromatography and Biological Recognition*, Chaiken, I. M., Parikh, I., and Wilchek, M., Eds., Academic Press, Orlando, Fla., 1983, 143.

25. **Taylor, H. C. and Chaiken, I. M.**, Active site ligand binding by an inactive ribonuclease S analogue, *J. Biol. Chem.*, 252, 6991, 1977.

26. **Dunn, B. M., DiBello, C., Kirk, K. L., Cohen, L. A., and Chaiken, I. M.**, Synthesis, purification, and properties of a semisynthetic ribonuclease S incorporating 4-Fluoro-L-Histidine at position 12, *J. Biol. Chem.*, 249, 6295, 1974.

27. **Taylor, H. C., Richardson, D. C., Richardson, J. S., Wlodawer, A., Komoriya, A., and Chaiken, I. M.**, "Active" conformation of an inactive semi-synthetic ribonuclease-S, *J. Mol. Biol.*, 149, 313, 1981.

28. **Bender, M. L. and Brubacher, L. J.**, *Catalysis and Enzyme Action*, McGraw-Hill, New York, 1973, 25.

29. **Winzor, D. J.**, Analytical gel chromatography, in *Physical Principles and Techniques of Protein Chemistry*, Part A, Leach, S. J., Ed., Academic Press, New York, 1969, Chap. 9.

30. **Kasai, K. and Ishii, S.**, Affinity chromatography of trypsin and related enzymes. V. Basic studies of quantitative affinity chromatography, *J. Biochem.*, 84, 1051, 1978.

31. **Kasai, K. and Ishii, S.**, Studies on the interaction of immobilized trypsin and specific ligands by quantitative affinity chromatography, *J. Biochem.*, 84, 1061, 1978.

32. **Wilkinson, G. N.**, Statistical estimations in enzyme kinetics, *Biochem. J.*, 80, 324, 1961.

33. **Eisenthal, R. and Cornish-Bowden, A.**, The direct linear plot. A new graphical procedure for estimating enzyme kinetic parameters, *Biochem. J.*, 139, 715, 1974.

34. **Steinhardt, J. and Beychok, S.**, Interactions of proteins with hydrogen ions and other small ions and molecules, in *The Proteins*, Vol. II, Neurath, H., Ed., Academic Press, New York, 1964, Chap. 8.

35. **Nichol, L. W., Ward, L. D., and Winzor, D. J.**, Multivalency of the partitioning species in quantitative affinity chromatography. Evaluation of the site-binding constant for the aldolase-phosphate interaction from studies with cellulose phosphate as the affinity matrix, *Biochemistry*, 20, 4850, 1981.

36. **Winzor, D. J., Ward, L. D., and Nichol, L. W.**, Quantitative considerations of the consequences of an interplay between ligand binding and reversible adsorption of a macromolecular solute, *J. Theor. Biol.*, 98, 171, 1982.

37. **Hogg, P. J. and Winzor, D. J.**, Quantitative affinity chromatography: further developments in the analysis of experimental results from column chromatography and partition equilibrium studies, *Arch. Biochem. Biophys.*, 234, 55, 1984.

38. **Denizot, F. C. and Delaage, M. A.**, Statistical theory of chromatography: new outlooks for affinity chromatography, *Proc. Natl. Acad. Sci. U.S.A.*, 72, 4840, 1975.

39. **Hethcote, H. W. and DeLisi, C.**, Non-equilibrium model of liquid column chromatography. I. Exact expressions for elution profile moments and relation to plate weight theory, *J. Chromatogr.*, 240, 269, 1982.

40. **Hethcote, H. W. and DeLisi, C.**, Non-equilibrium model of liquid column chromatography. II. Explicit solutions and non-ideal conditions, *J. Chromatogr.*, 240, 282, 1982.

41. **Hethcote, H. W. and DeLisi, C.**, Determination of equilibrium and rate constants by affinity chromatography, *J. Chromatogr.*, 248, 183, 1982.

42. **Hethcote, H. W. and DeLisi, C.**, Quantitative affinity chromatography: new methods for kinetic and thermodynamic characterization of macromolecular interactions, in *Affinity Chromatography and Biological Recognition*, Chaiken, I. M., Parikh, I., and Wilchek, M., Eds., Academic Press, Orlando, 1983, 119.

43. **Chaiken, I. M.**, Preparative and analytical affinity chromatography of neurophysins on methionyl-tyrosyl-phenylalanyl-aminohexyl-agarose, *Anal. Biochem.*, 97, 302, 1979.

44. **Angal, S. and Chaiken, I. M.**, Interdependence of neurophysin self-association and neuropeptide hormone binding as expressed by quantitative affinity chromatography, *Biochemistry*, 21, 1574, 1982.

45. **Chaiken, I. M., Tamaoki, H., Brownstein, M. J., and Gainer, H.**, Onset of neurophysin self-association upon neurophysin/neuropeptide hormone precursor biosynthesis, *FEBS Lett.*, 164, 361, 1983.

46. **Abercrombie, D. M., Kanmera, T., Angal, S., Tamaoki, H., and Chaiken, I. M.**, Cooperative interactions in neurophysin-neuropeptide hormone complexes, *Int. J. Pept. Prot. Res.*, 24, 218, 1984.

47. **Swaisgood, H. S. and Chaiken, I. M.**, Characterization of the neurophysin-neuropeptide hormone system by analytical high performance affinity chromatography, *J. Chromatogr.*, 327, 193, 1985.

48. **Kanmera, T. and Chaiken, I. M.**, Molecular Properties of the oxytocin/bovine neurophysin biosynthetic precursor: studies using a semisynthetic precursor, *J. Biol. Chem.*, in press.

49. **Richter, D.**, Vasopressin and oxytocin are expressed as polyproteins, *Trends Biochem. Sci.*, 8, 278, 1983.

50. **Brownstein, M. J., Russell, J. T., and Gainer, H.**, Synthesis, transport, and release of posterior pituitary hormones, *Science*, 207, 373, 1980.

51. **Pickering, B. T.**, The neurosecretory neurone: A model system for the study of secretion. *Essays Biochem.*, 14, 45, 1978.

52. **Chaiken, I. M., Fischer, E. A., Giudice, L. C., and Hough, C. J.**, *In vitro* synthesis of hypothalamic neurophysin precursors, in Hormonally Active Brain Peptides, McKerns, K. W. and Pantic, V., Eds., Plenum Press, New York, 1982, 327.

53. **Robinson, A. G., Verbalis, J. G., Amico, J. A., and Seif, S. M.**, Recent advances in neurohypophysial research, *Int. Rev. Physiol.*, 24, 1, 1981.

54. **Soloff, M. S. and Pearlmutter, A. F.**, Biochemical actions of neurohypophysial hormones and neurophysins, in *Biochemical Actions of Hormones*, Vol. VI, Litwack, G., Ed., Academic Press, New York, 1979, 263.

55. **Acher, R.**, Neurophysins: molecular and cellular aspects, *Angew. Chem. Int. Ed. Eng.*, 18, 846, 1979.

56. **Robinson, I. C. A. F., Edgar, D. H., and Walter, J. M.**, A new method of coupling [8-Lysine]vasopressin to agarose: the purification of neurophysins by affinity chromatography, *Neurosciences*, 1, 35, 1976.

57. **Chaiken, I. M., Taylor, H. C., and Randolph, R. E.**, Conformation effects associated with the interaction of polypeptide ligands with neurophysins, *N.Y. Acad. Sci.*, 248, 442, 1975.

58. **Menendez-Botet, C. J. and Breslow, E.**, Chemical and physical properties of the disulfides of bovine neurophysin-II, *Biochemistry*, 14, 3825, 1975.

59. **Chaiken, I. M., Kanmera, T., and Sequeira, R. P.**, Folding and enzymatic processing of precursors of biologically active peptides and proteins, in *Opioid Peptides: Molecular Pharmacology, Biosynthesis, and Analysis*, Rapaka, R. S. and Hawks, R. L., Eds., NIDA Press, 1986, 3.

Chapter 3

PRACTICAL APPROACHES FOR THE MEASUREMENT OF RATE CONSTANTS BY AFFINITY CHROMATOGRAPHY

Rodney R. Walters

TABLE OF CONTENTS

I. INTRODUCTION

The measurement of the equilibrium-binding constant between an immobilized-affinity ligand and a solute (or analyte or ligate) is relatively straightforward because the biospecific interaction of interest is almost always the major cause of retention in the affinity-chromatographic column. However, the measurement of the rate of association or dissociation of the same complex is not as straightforward because many kinetic processes occur in the column. The adsorption or desorption rate can only be measured if it is the rate-limiting process. Thus, in many cases it may be nearly impossible to determine the desired kinetic parameters or, alternatively, the data may be misinterpreted and false conclusions regarding the rate constants may be made. It is thus imperative that these other kinetic processes be quantitatively assessed so as to ensure the accuracy of any conclusions.

In the chapter by DeLisi and Hethcote, the theory of rate-constant measurement under isocratic-elution conditions was discussed. In this chapter the isocratic method and two other methods are discussed from a more practical point of view using experimental data, example calculations, and computer simulations.

The theory used here is cast in a somewhat different form than in previous chapters. Terms such as the capacity factor (k′) and plate height (H) that are widely used in the chromatographic literature will be used in this discussion. The glossary at the end of the chapter summarizes these terms and also lists the abbreviations used in previous chapters. Although terms such as the "capacity factor" may seem confusing to those unfamiliar with chromatographic theory, in fact the opposite is true. For example, the retention time (t_r) of a solute depends on many factors including the column dimensions and flowrate (F); however, k′ is independent of these variables. Therefore, if one sets the condition k′ $\geq$ 10, then one can easily interpret what this means for any column operated under any conditions. In contrast, the condition t_r $\geq$ 5 min applies to only a particular column under particular conditions.

The purpose of the theory and computer simulation sections is to help define the limitations of each of the methods. These limitations are then used in the example calculations to determine the necessary experimental conditions and whether the desired measurement can even be made. The calculations are shown in detail so that a scientist who is not an expert in chromatographic theory can duplicate them for other biochemical systems of interest.

The importance of preliminary calculations cannot be overemphasized. There are many practical as well as theoretical limitations in equilibrium and rate-constant measurements. Many of these are not apparent until the calculations are performed.

II. REVIEW OF CHROMATOGRAPHIC BAND-BROADENING THEORY

The most well-known of the many works pertaining to chromatographic band-broadening is the text by J. C. Giddings entitled "Dynamics of Chromatography".[1] Other useful works include references 2 to 10. Hethcote and DeLisi[11-14] have extended the theory to several cases of great importance in affinity chromatography, e.g., "normal-role" affinity chromatography in which the kinetics of inhibitor-solute binding enter into the equations.[13]

Although chromatographic theory is well-developed, there is some question as to how accurately real columns are described. This problem is related to two main factors:

1. The complexities of the flow profile in a packed bed
2. The heterogeneities always present in real columns, e.g., particle-size distribution and irregular packing

As a result, it may never be possible to exactly describe the kinetic processes in a packed column.

The major processes that are known to cause broadening of an initially narrow band of solute as it passes through a column are the following. In the interparticle space (excluded volume) there will generally be a parabolic-flow profile, just as in a capillary tube, with the velocity being faster in the middle of the channel than at the walls. Band-broadening will be opposed by diffusion from the fast flowstreams into the slow flowstreams and vice versa; this process is called mobile-phase mass-transfer. Also, in the interparticle space, the flow-paths followed by some solute molecules will be longer and more tortuous than others, leading to the band-broadening process called eddy diffusion. A third process in the inter-particle space is longitudinal or axial diffusion, the spreading of the solute band in the direction of flow due to diffusion. In liquid-mobile phases, this process is usually negligible.

In the intraparticle space (pore volume) the mobile phase is stagnant, i.e., not flowing. Slow equilibration of solute molecules between the stagnant- and flowing-mobile phase results in a stagnant-mobile phase mass-transfer process that is opposed by diffusion.

When solute molecules reach the surface of the stationary phase, adsorption and desorption may take place. Slow adsorption or desorption kinetics lead to a kinetic- or stationary-phase mass-transfer process. (Of course, all the band-broadening is of kinetic origin, however, in this chapter "kinetic" will refer to only slow-adsorption/desorption mass-transfer processes and "diffusional" will refer to stagnant-mobile phase mass-transfer processes.)

An illustration of the various band-broadening processes can be found in Figure 2.3 of Reference 10.

Mathematically, the effect of these processes on the width of the solute peak when it leaves the column is expressed in terms of the standard deviation (σ), variance (σ^2), or plate height (H) of the eluted peak under isocratic (constant-mobile phase composition), small-zone (the sample is applied to the column as a band of volume negligible compared to the volume when it leaves the column), and linear elution (the number of molecules of solute applied is negligible compared to the number of molecules of stationary phase in the column) conditions.

The plate height is calculated from the peak variance, retention time (t_r), and column length (L_{col}) either directly or by first calculating the number of theoretical plates (N) in the column:

$$N = \frac{t_r^2}{\sigma^2} \tag{1}$$

$$H = \frac{L_{col}}{N} \tag{2}$$

(Note: σ^2 must be expressed in the same units as t_r^2.)

As will become clear later, H is frequently a more useful parameter to use than is σ^2. For both parameters the individual contributions to band-broadening are additive with the exception of the eddy diffusion and mobile-phase mass-transfer terms which are thought to be coupled and thus combined into one term (H_m or σ^2_m):

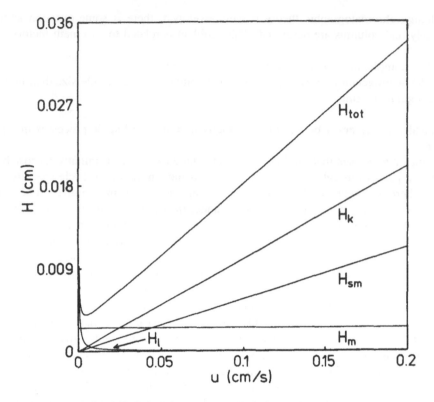

FIGURE 1.   Theoretical plots of the various band-broadening contributions to plate height as a function of mobile-phase linear velocity according to the van Deemter Equation. In liquid-mobile phases, one seldom uses velocities near or below the minimum in the curve; hence, H_l is normally negligible. Conditions: $k' = 1$, $k_{-3} = 5 \text{ sec}^{-1}$, $k_{-1} = 40 \text{ sec}^{-1}$, $D_m = 1.67 \times 10^6 \text{ cm}^2/\text{sec}$, $d_p = 5 \times 10^{-4} \text{ cm}$, $V_e = V_p$.

$$H_{tot} = H_m + H_l + H_{sm} + H_k \tag{3}$$

$$\sigma_{tot}^2 = \sigma_m^2 + \sigma_l^2 + \sigma_{sm}^2 + \sigma_k^2 \tag{4}$$

The subscripts refer to the total (tot), mobile phase (m), longitudinal diffusion (l), stagnant mobile phase (sm), and adsorption/desorption kinetic (k) contributions to band-broadening.

In addition, there may be significant extra-column volume and band-broadening due to connecting tubing, detector volume, injection-loop volume, etc. The raw values of t_r and σ^2 should first be corrected for these contributions.

If one wishes to measure a dissociation-rate constant, then two criteria must be met in addition to the conditions listed above:

1. H_k must be significant compared to H_{tot} (Preferably, $H_k \geq 0.9 \, H_{tot}$)
2.   The other band-broadening terms must be independently measured or calculated or be negligible

It is apparent that one must know the exact mathematical dependence of each band-broadening term on the various physical properties of the column. Many of these dependencies have been determined both experimentally and theoretically. For example, the dependence of H_{tot} on linear velocity (u) or flowrate (F) is shown in Figure 1. H_l is inversely proportional to u while H_{sm} and H_k are proportional to u. The dependence of H_m on u is still controversial.[7,8] According to the van Deemter Equation,[2] H_m is independent of u and according to the Knox Equation,[3] H_m is a function of $u^{1/3}$.

One of the greatest uncertainties is exactly how each term changes with retention. Recent work has shown that none of the theoretical relationships accurately describe H_{tot} as a function of retention time or capacity factor.[6,7]

Another problem is the presence in the theoretical equations of various parameters which are difficult to determine. For example, the H_m term of Horvath and Lin[5] contains "structural parameters" λ and ω whose values are not easily determined and which vary from column to column.

It is clear that there is still much to learn concerning band-broadening in packed columns. To utilize the current theory for the measurement of rate constants, one must make several simplifying assumptions. Insofar as possible, these assumptions should be tested experimentally to make sure that they are reasonable.

The theory developed by Hethcote and DeLisi[11-14] describes only the H_{sm} and H_k band-broadening terms. While it is reasonable to ignore H_l, H_m is usually not negligible. Ignoring the eddy diffusion-mobile phase mass-transfer term can lead to false conclusions. For example, one might conclude that nonporous beads would be ideal for making kinetic measurements because $H_{sm} = 0$, hence $H_{tot} = H_k$; however, this is a false conclusion because even a nonretained solute will undergo band-broadening (due to the eddy-diffusion and mobile-phase mass-transfer processes) in a column of nonporous beads. Ignoring H_m or other band-broadening terms will lead to overestimation of the magnitude of H_k and thus to underestimation of the dissociation-rate constant.

III. ISOCRATIC METHOD

A. Theory

To determine rate constants under isocratic conditions, we have utilized the stagnant-mobile phase and kinetic terms derived by Hethcote and DeLisi[11-14], but modified to contain an eddy diffusion-mobile phase mass-transfer term as in the van Deemter Equation.[2]

According to Hethcote and DeLisi,[11-14] the two major band-broadening processes can be written as

$$E_e \overset{k_1}{\underset{k_{-1}}{\rightleftharpoons}} E_p \tag{5}$$

$$E_p + L \overset{k_3}{\underset{k_{-3}}{\rightleftharpoons}} E_p - L \tag{6}$$

where k_1 and k_{-1} are the first-order constants for diffusion of a solute, E, into the pore volume, V_p, or back into the excluded volume, V_e; k_3 is the second order adsorption-rate constant for the formation of a complex between the solute and an immobilized-affinity ligand, L; and k_{-3} is the first-order rate constant for dissociation of the complex.

A competing ligand or inhibitor, I, must generally be added to the mobile phase to decrease the retention of E. In the case of "normal-role" affinity chromatography, the inhibitor binds to the solute with rate constants k_2 and k_{-2}

$$E + I \overset{k_2}{\underset{k_{-2}}{\rightleftharpoons}} E - I \tag{7}$$

In the case of "reversed-role" affinity chromatography, the inhibitor binds to the ligand

$$L + I \underset{k_{-2}}{\overset{k_2}{\rightleftharpoons}} L - I \tag{8}$$

In both modes, the inhibitor has the same effect on retention. However, the modes have different expressions for band-broadening.

The equilibrium-binding constants can be written for each process:

$$\frac{V_p}{V_e} = \frac{k_1}{k_{-1}} \tag{9}$$

$$K_2 = \frac{k_2}{k_{-2}} \tag{10}$$

$$K_3 = \frac{k_3}{k_{-3}} = \frac{\{E - L\}}{[E]\{L\}} \tag{11}$$

In Equation 11, the { } represent surface concentrations. Note that the units of K_3 are the same as for a solution-binding constant (M^{-1}), but because the ligand is attached to a surface, one might not expect the equilibrium and rate constants to be the same as in solution. Equating the solution K_3 with the chromatographic K_3 means that the "effective" molar concentration of ligand in the column is m_L/V_p where m_L is the total number of moles of ligand in the column. In fact, the generally good agreement between solution and chromatographic K_3 values suggests that this is a good approximation; however, this does not ensure that the rate constants are the same in both cases.

The equations of Hethcote and DeLisi[11-14] were derived assuming certain conditions described in Part II. Isocratic conditions mean that the concentration of I in the mobile phase is constant during a particular chromatographic run. Linear-elution conditions mean that the adsorption process is first order, which in turn means that so little solute is applied to the column that an insignificant number of ligand sites are occupied by E and thus {L} in Equation 11 is constant and equal to m_L/A where A is the total surface area of the matrix. The first-order adsorption-rate constant, k_3^*, can be written as

$$k_3^* = \frac{k_3 m_L}{V_p} \tag{12}$$

1. Retention Theory

The retention time, t_r, for a monovalent solute, given by Equations 130 and 217 in the DeLisi and Hethcote chapter, is rewritten here as

$$t_r = \frac{L_{col}}{u_e} \left(1 + \frac{V_p}{V_e} \left(1 + \frac{K_3 m_L/V_p}{1 + K_2[I]} \right) \right) \tag{13}$$

where u_e is the linear velocity of an excluded solute.

The capacity factor, k', is calculated by

$$k' = \frac{t_r - t_m}{t_m} \tag{14}$$

where t_m is the retention time of a nonretained solute. The capacity factor can be thought of as the number of multiples of the column-void time or volume that the solute spends in the stationary phase.

Using some of the definitions in the glossary, Equations 13 and 14 are combined to give

$$k' = \frac{K_3 m_L}{V_m(1 + K_2[I])} \tag{15}$$

where V_m is the total volume of mobile phase in the column. It should be noted that if the solute, E, is partially excluded from the pores of the matrix, then V_m, V_p, u, k', etc. should all be measured using the partially excluded solute.

The reciprocal of Equation 15 is equivalent to Equation 11 in the Swaisgood and Chaiken chapter and shows that a plot of $1/k'$ vs. [I] will be linear with slope $V_m K_2/K_3 m_L$ and intercept $V_m/K_3 m_L$ and from which equilibrium constants K_2 and K_3 may be calculated.

2. Band-Broadening Theory

For normal-role affinity chromatography, the kinetic- or stationary-phase contribution to the variance is given by Equation 147 in the DeLisi and Hethcote chapter. It is rewritten here without the exponential term, which in many cases is negligible

$$\sigma_k^2 = \frac{2 L_{col} K_3 m_L}{u_e V_e(1 + K_2[I])} \left(\frac{1}{k_{-3}} + \frac{K_2[I] K_3 m_L}{(1 + V_p/V_e) V_e k_{-2}(1 + K_2[I])^2} \right) \tag{16}$$

Substituting as before and rewriting in terms of k' yields

$$\sigma_k^2 = 2k' t_m \left(\frac{1}{k_{-3}} + \frac{k' K_2[I]}{k_{-2}(1 + K_2[I])} \right) \tag{17}$$

From Equations 1, 2, 14, and 17, the kinetic-plate height is given by

$$H_k = \frac{L_{col} \sigma_k^2}{t_r^2} = \frac{L_{col} \sigma_k^2}{t_m^2(1 + k')^2}$$

$$= \frac{2uk'}{(1 + k')^2} \left(\frac{1}{k_{-3}} + \frac{k' K_2[I]}{k_{-2}(1 + K_2[I])} \right) \tag{18}$$

In the case of reversed-role affinity chromatography, the term containing k_{-2} is not needed (see Equation 219 in the DeLisi and Hethcote chapter)

$$H_k = \frac{2 uk'}{k_{-3}(1 + k')^2} = \frac{2 u_e V_e k'}{k_{-3} V_m(1 + k')^2} \tag{19}$$

This equation has been derived by many others previously.[1,5] Because of the complexity of the normal-role case (Equation 18), in this chapter only the simpler reversed-role case (Equation 19) will be examined in detail.

The stagnant-mobile phase mass-transfer contribution to the peak variance is given by Equations 136 and 219 in the DeLisi and Hethcote chapter, rewritten here as

$$\sigma_{sm}^2 = \frac{2 V_p L_{col}}{V_e u_e k_{-1}} \left(1 + \frac{K_3 m_L}{V_p(1 + K_2[I])} \right)^2 \tag{20}$$

After substitution, this yields

$$\sigma^2_{sm} = \frac{2V_p t_m}{V_m k_{-1}} \left(1 + \frac{V_m k'}{V_p} \right)^2 \tag{21}$$

In analogy to Equation 18, the stagnant-mobile phase plate-height is given by

$$H_{sm} = \frac{2uV_p(1 + V_m k'/V_p)^2}{V_m k_{-1}(1 + k')^2}$$

$$= \frac{2u_e V_e V_p(1 + V_m k'/V_p)^2}{V_m^2 k_{-1}(1 + k')^2} \tag{22}$$

This equation is identical in form to that of Horvath and Lin,[5] but does not indicate the physical dependence of k_{-1}. By comparing the two equations

$$k_{-1} = \frac{60 \gamma' D_m}{d_p^2} \tag{23}$$

where D_m is the diffusion coefficient of the solute, γ is a tortuosity factor which corrects for the slower diffusion inside the pores of the matrix, and d_p is the particle diameter of the matrix. The particles are assumed to be spherical and totally porous.

The value of γ is usually assumed to be approximately 0.5;[5,9] however, if the size of the solute approaches the pore diameter, there is evidence to indicate that γ decreases further.[15] A more conservative estimate of γ to use in calculations would be 0.1.

The remaining plate-height term is the eddy diffusion-mobile phase mass-transfer term, H_m, which will be assumed here to be independent of, "u" and k' and of a magnitude to be determined experimentally. In well-packed columns, H_m should be no more than twice the particle diameter.[7] Columns that are packed in the research laboratory will probably not be this efficient, so a more conservative estimate of H_m for calculation purposes would be 5 d_p.

The variance due to the eddy diffusion-mobile phase mass-transfer process is given by

$$\sigma^2_m = \frac{t_m^2(1 + k')^2 H_m}{L_{col}} \tag{24}$$

3. Plate Height vs. Variance

Although the same results should be obtained whether the equations for variances or plate heights are used, there are some practical advantages for using plate heights. First, when a column is operated near the "optimum" velocity (minimum of Figure 1), $H_{tot} \sim H_m \sim 2$ d_p. Thus, plate height can be thought of in terms of multiples of the particle diameter, and it is easy to estimate whether a given peak is unusually broad or whether a column is properly packed. Since H is independent of column dimensions it is also easy to compare various columns.

A second advantage of plate heights is revealed by Figure 2. The plot of H_k vs. k' exhibits a maximum at k' = 1, whereas the plot of H_{sm} vs. k' plateaus at large k'. Thus, a maximum in an experimental plot of H_{tot} vs. k' is good qualitative evidence that slow-adsorption/ desorption kinetics contribute significantly to the plate height and thus that it will be possible to measure k_{-3}. The variance plots (Figure 2b) do not have this useful feature.

It is also apparent from both the plate height and variance plots in Figure 2 that the most precise measurements of k_{-3} will be obtained when k' $\sim$ 1 since H_k and σ^2_k are largest relative to H_{sm} and σ^2_{sm} at this point.

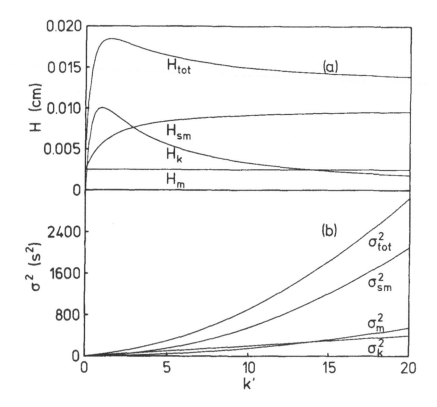

FIGURE 2. Theoretical plots of the various contributions to plate height (a) and variance (b) as a function of capacity factor. Conditions: same as Figure 1 except u = 0.1 cm/sec and L_{col} = 5 cm.

B. Computer Simulations

Computer-generated peak profiles are useful for predicting peak shapes and other parameters which cannot be calculated directly from theory. We have utilized a computer model which operates in close analogy to a real column. The imaginary column is divided lengthwise into a series of segments. Each segment is in turn divided into excluded volume, pore volume, and stationary phase (Figure 3). Utilizing the solution to the two-step reversible-rate equation

$$E_e \underset{k_{-1}}{\overset{k_1}{\rightleftharpoons}} E_p \underset{k_{-3}}{\overset{k_3^*}{\rightleftharpoons}} E - L \tag{25}$$

given in Reference 16 and written in terms of k', V_p/V_e, k_{-1}, and k_{-3}, each iteration of the computer consists of moving the solute in the excluded volume ahead one segment, then equilibrating it between the three parts of each segment. The program yields accurate values of the plate height as long as the rate constants are less than 0.1 iterations^{-1} so that the system is far away from equilibrium. The program models only the band-broadening due to the H_{sm} and H_k terms.

This program has been used to determine:

1. How many theoretical plates should be generated in an experimental system
2. How to measure the retention time and variance
3. The effect of some types of nonideal behavior on the results

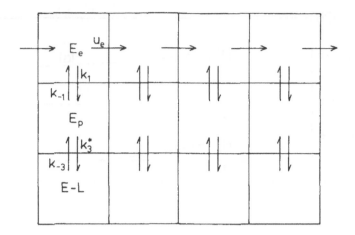

FIGURE 3.   Illustration of the model used for computer simulations. From top to bottom, the rows represent the moving-mobile phase, stagnant-mobile phase, and stationary phase. The flow is from left to right along the length of the column.

Simulations performed using a wide range of rate constant, V_p/V_e, and k' values have shown that roughly 50 or more theoretical plates need to be generated in the column to produce peaks that are moderately symmetric and free of the peculiar "split-peak" behavior that will be discussed in the next section.

The question of how to measure the retention time and variance from the peak profile is, from a theoretical standpoint, quite clear. The theoretical equations of Hethcote and DeLisi and others are all based on the statistical moments of the peaks. The zeroth moment is the peak area, the first moment (or center of gravity) is the retention time, and the second moment is the variance. Using a computer for data acquisition and calculation, the statistical moments are calculated by the summation method using the individual data points (i), peak heights (h(i)), and time interval between points (Δt)

$$\text{Zeroth moment} = M_o = \Delta t \cdot \Sigma h(i) \tag{26}$$

$$\text{First moment} = t_r = \frac{(\Delta t)^2 \cdot \Sigma(i \cdot h(i))}{M_o} \tag{27}$$

$$\text{Second moment} = \sigma^2 = \frac{(\Delta t)^3 \cdot \Sigma(i^2 \cdot h(i))}{M_o} - t_r^2 \tag{28}$$

Although this calculation is simple to perform on a computer, the practical problem is to determine when the peak height has decreased to zero at the end of the peak tail. For asymmetric, tailed peaks, small errors in determining the baseline can lead to estimates of variance that are as little as one-half of the true value.[17] This is due to the sensitivity of the second moment to the area under the tail of the peak.

Another approach is to assume some model-peak shape, e.g., an exponentially modified Gaussian peak.[17] This approach helps to minimize the errors caused by the peak tail, but its accuracy depends on the peak having the expected shape.

A third approach is to assume that the peak is close to Gaussian in shape. The variance can then be determined from the width-at-half-height, $w_{1/2}$ (Figure 4)

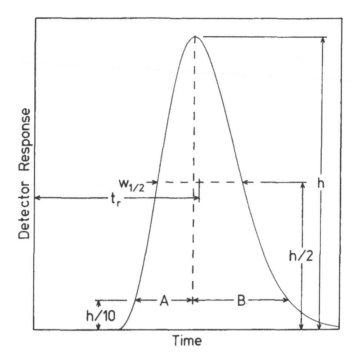

FIGURE 4. A computer-simulated peak showing the width-at-half-height ($w_{1/2}$), the peak-center-at-half-height (t_r), and the peak asymmetry (B/A).

$$\sigma^2 = w_{1/2}^2/8 \ln 2 \tag{29}$$

A convenient measure of the retention time is the center of the peak at half-height (Figure 4). If the peaks are badly tailed, the error due to the use of this Gaussian-approximation method will be even greater than the error usually associated with baseline uncertainty in the summation method.[17]

As a general rule for real tailed peaks, the Gaussian approximation yields inaccurate but precise estimates of the moments, while the summations methods yield imprecise but accurate moments. To help determine which method to use, peak profiles were computer-simulated as above. The peak-asymmetry ratio (B/A, Figure 4) and the moments determined by the summation and Gaussian approximation methods were determined. For $N \geqslant 50$, the errors involved in using the Gaussian approximation were only a few percent because the peaks proved to be quite symmetric.

However, experimental data are always more tailed than the simulated data.[18] Computer simulations indicate that this could be due to particle size heterogeneity or other such nonideal behavior.[18] The Gaussian-approximation method is less accurate as the peak asymmetry increases, but does give a slightly more accurate estimate of the median value of the rate constant than the summations method.[18]

Therefore, we have concluded that the better precision of the Gaussian-approximation method combined with its smaller sensitivity to heterogeneity makes it the method of choice for kinetic studies. However, it should be pointed out that this choice may bias the results such that the measured rate constants could be too high by as much as a factor of two.

Another interesting result of the computer-simulation studies and the theory is that the peak asymmetry should be independent of k' if $H_{sm} \gg H_k$, but the peak asymmetry will decrease as k' increases if $H_k \gg H_{sm}$.[18] This provides another way (in addition to the shape of plots of H_{tot} vs. k') to determine whether H_{sm} or H_k is the dominant band-broadening term.

C. Example Calculations

To perform preliminary calculations, it is helpful to use several approximations. To calculate H_m, it was pointed out earlier that one might expect $H_m \leq 5$ dp; to calculate H_{sm}, it was also pointed out that $\gamma \sim 0.1$ would be a reasonable estimate. For many matrices if the solute is able to penetrate all of the pore volume, then $V_p = V_e = V_m/2$ is a good estimate. A column packed with spherical particles should have $V_e/V_{col} \sim 0.4;$[5,9] this provides a simple way to calculate V_m, V_p, and V_e from the column dimensions.

A good choice for k' in preliminary calculations is $k' = 1$ since this is the best point to measure k_{-3}. From k' one can calculate the inhibitor concentration (Equation 15), if an estimate of m_L/V_m and some guess as to the values of K_2 and K_3 are made. An estimate of m_L/V_m can be made by assuming, e.g., that it is possible to immobilize the affinity ligand such that the active ligand corresponds to 1/10 of a monolayer. The size of the ligand can be estimated from the Stokes radius using the diffusion coefficient or molecular weight. The surface area of the matrix can be estimated from the packing density (~ 0.4 g/mℓ for silica) and the data supplied by the manufacturer (e.g., 50 to 100 m^2/g for 300 Å pore-size silica). As an example, m_L/V_m for albumin (diameter $= 74$ Å) on a 50 m^2/g matrix is calculated as follows:

$$\frac{\text{moles albumin}}{\text{m}^2} = \frac{1 \text{ molecule}}{(74 \times 10^{-10}\text{m})^2} \times \frac{1 \text{ mole}}{6 \times 10^{23} \text{ molecules}} = 3.0 \times 10^{-8} \text{ mol/m}^2$$

$$\frac{\text{column surface area}}{V_m} = \frac{50 \text{ m}^2}{1 \text{ g silica}} \times \frac{0.4 \text{ g silica}}{1 \text{ m}\ell \text{ empty column}} \times \frac{1 \text{ m}\ell \text{ empty column}}{0.8 \text{ m}\ell \ V_m} = 25 \text{ m}^2/\text{m}\ell$$

$$\frac{m_L}{V_m} \text{ (1/10 monolayer)} = 0.1 \times 3.0 \times 10^{-8} \text{ mol/m}^2 \times 25 \text{ m}^2/\text{m}\ell = 75 \frac{\text{nmol}}{\text{m}\ell}$$

The column diameter is usually 0.41 or 0.46 cm. At a flowrate of 1 mℓ/min (a typical value for silica matrices), the linear velocity will be given by

$$u = \frac{L}{t_m} = \frac{L}{\pi r^2 L \times 0.8/F} = \frac{1}{\pi(0.23\text{cm})^2 \times 0.8/0.0167 \text{ m}\ell/\text{sec}} = 0.13 \text{ cm/sec}$$

The factor 0.8 is the ratio V_m/V_{col}.

To calculate k_{-1} using Equation 23, D_m can be estimated from the solute molecular weight.[19] H_{sm} can then be estimated. A guess as to the value of k_{-3} allows one to calculate H_k and finally H_{tot}. The column length can then be chosen using H_{tot} and Equation 2 so that $N \geq 50$. If the length is unreasonably large one can use a smaller linear velocity and recalculate H_{sm} and H_k.

Detection problems can be examined by assuming $m_E \sim 0.01 \ m_L$ so that linear-elution conditions are maintained. Now that the column length is known, one can calculate m_L and t_m. Using k' and t_m, t_r can be calculated (Equation 14). Using N and t_r, σ can be calculated (Equation 1). The average solute concentration in the detector will be roughly $m_E/4\sigma F$.

One can also calculate whether the extracolumn variance introduced by the injector, connecting tubing, and detector is too large. If the extracolumn band-broadening is to be less than 10% of the column band-broadening (or more preferably, 1%), then a rough rule to follow is

$$V_{inj}^2 + V_{det}^2 + V_{tub}^2 \leq 0.1 \ \sigma^2 F^2 \tag{30}$$

where V is the volume of each component. Many detectors contain a length of heat-exchanger tubing; hence a 10 $\mu\ell$ flow cell might often be connected to a 40 $\mu\ell$ heat-exchanger. The heat-exchanger can usually be removed if it is necessary to decrease extracolumn band-broadening.

1. DNP-Antibody Kinetics

As a first example, consider the interaction of the immobilized-antibody MOPC 315 with one dinitrophenyl hapten used as the solute and another used as the inhibitor. The solution kinetics for this system are known to be very rapid[20]

$$k_{-3} \sim 500 \text{ sec}^{-1}$$

$$K_3 \sim 10^6 \ M^{-1}$$

$$K_2 \sim 10^6 \ M^{-1}$$

Assume that $D_m \sim 10^{-5} \text{cm}^2/\text{s}$ and $d_p = 5$ μm in addition to the earlier assumptions that $V_p = V_e$, $k' = 1$, and $u = 0.13$ cm/s. The plate heights are calculated to be:

$$H_m = 2.5 \times 10^{-3} \text{ cm}$$

$$k_{-1} = 240 \text{ sec}^{-1}$$

$$H_{sm} = 1.2 \times 10^{-3} \text{ cm}$$

$$H_k = 1.3 \times 10^{-4} \text{ cm}$$

It is apparent that H_k constitutes only about 3% of the total plate-height, hence it is not possible to measure k_{-3} under these conditions. It should be noted that these are fairly favorable conditions because a small, fast-diffusing solute and a very-small matrix diameter were chosen. Nevertheless, these conditions limit one to measuring k_{-3} values less than about 1% of the above value.

2. Concanavalin A — Sugar Kinetics

A similar system but with slower kinetics is one in which concanavalin A (Con A) is immobilized and a sugar is used as the solute (e.g., p-nitrophenyl α-D-mannopyranoside, PNPM) and as the inhibitor (e.g., methyl α-D-mannopyranoside, MDM). The solution data is as follows[21]

$$k_{-3} = 6.2 \text{ sec}^{-1}$$

$$K_3 = 8.7 \times 10^3 \ M^{-1}$$

$$K_2 = 3.3 \times 10^3 \ M^{-1}$$

Using the same conditions as before

$$H_m = 2.5 \times 10^{-3} \text{ cm}$$

$$k_{-1} = 240 \text{ sec}^{-1}$$

$$H_{sm} = 1.2 \times 10^{-3} \text{ cm}$$

$$H_k = 1.1 \times 10^{-2} \text{ cm}$$

Thus, H_k is about 70% of the total plate height and it should be possible to determine k_{-3}.

One can now calculate other useful experimental parameters as described above. First, assuming 1/10 monolayer of active-Con A, a 50 m²/g surface-area matrix, the same ligand diameter as for the albumin calculation, and two binding sites/Con A molecule, m_L/V_m = 150 nmol/mℓ. From Equation 15, [I] = 9×10^{-5} M. Using the total plate height (0.014 cm), a column length of 5 cm would generate 350 plates, which more than meets the requirement of about 50 plates. If the column diameter is 0.46 cm, V_{col} = 0.83 mℓ, V_m = 0.66 mℓ, t_m = 40 sec, t_r = 80 sec, and m_L = 100 nmol. About 1 nmol PNPM should be applied to the column. From Equation 1, σ = 4.3 sec and the average concentration of solute in the detector is 4×10^{-6} M. This would be near the lower limit of detection of an absorbance detector.

Since σ = 4.3 sec, it is necessary to ensure that the detector time constant is small so that the peak will not be broadened by the electronic damping. Since σF = 72 $\mu\ell$ it is important that low-volume connecting tubing and a small injection-loop volume and detector-cell volume be used. If V_{inj} = V_{det} and V_{tub} ~ 0, the maximum-injection loop and detector-cell volume is calculated to be 16 $\mu\ell$ from Equation 30.

All of these calculations suggest that there should not be any major practical problems associated with the determination of k_{-3} for this system.

It is apparent that one would normally use the smallest-diameter matrix available to maximize k_{-1} and minimize H_m and H_{sm}. In the above example, if d_p were increased from 5 μm to 100 μm (a typical diameter for agarose), then H_{sm} would increase 400-fold and H_m would increase 20-fold; H_k would only represent 2% of the total band-broadening and k_{-3} could not be determined.

3. Protein A — Immunoglobulin G

In general, one would like to apply the isocratic method to macromolecule-macromolecule interactions as well as to macromolecule-small solute interactions. The two major problems are: (1) the slower diffusion rate of macromolecular solutes and (2) the difficulty in finding a suitable competing inhibitor.

With currently-available matrices, the dissociation-rate constants of macromolecules can be determined if they are less than about 10^{-1} to 10^{-2} sec^{-1}. This does seem to be the case for many strong antibody-antigen complexes.[20] Here we will consider the similar case of the protein A-immunoglobulin G (IgG) interaction in which the protein A is immobilized and the IgG is the solute. There is not much choice of competing inhibitor except IgG. Assume that the solute IgG is labeled so that it can be selectively detected. Some preliminary work[22,23] has indicated the following

$$k_{-3} \sim 10^{-3} \text{ sec}^{-1}$$

$$K_2 = K_3 = 4 \times 10^8 \ M^{-1}$$

The diffusion coefficient of IgG is 4×10^{-7} cm²/sec; assume a 10 μm diameter matrix is used. From Equation 23, k_{-1} = 2.4 sec^{-1}. If, as before, a flow rate of 1 mℓ/min is chosen, calculations readily indicate that the column would have to be quite long to generate 50 plates. Instead, choose u such that 50 plates are generated in a 10 cm-long column, i.e., H_{tot} = 0.2 cm. If H_m = 5×10^{-3} cm, then H_{sm} + H_k = 0.195 cm. Setting V_e = V_p and k' = 1, one can calculate from Equations 19 and 22 that u = 3.9×10^{-4} cm/sec, H_{sm} = 4×10^{-4} cm, and H_k = 0.195 cm. Thus, H_k should totally dominate the band-broadening. Also, t_m = 7.1 hr and t_r = 14.2 hr. This will not be a fast experiment!

If the protein A were assumed to be a 1/10 monolayer as before, it turns out that the inhibitor concentration, while not large on a molar basis, is quite high on a weight/volume basis. Therefore, assume $[I] = 1$ mg IgG/m$\ell = 6.7 \times 10^{-6} M$ and calculate from Equation 15 that $m_L/V_m = 6.7$ nmol/mℓ. If the column is 10 cm $\times$ 0.46 cm, $V_m = 1.3$ mℓ, so $m_L = 9.0$ nmol. The amount of IgG applied as solute would be about 0.1 nmol. From V_m and t_m, the flowrate is calculated to be 3×10^{-3} mℓ/min. From Equation 1, $\sigma = 2.0$ hr and the average concentration of solute in the detector is $7 \times 10^{-8} M$. To detect this low concentration, a fluorescent or radioactive label on the IgG would be required.

So while it should, in principle, be easy to measure k_{-3} for this example, the practical limitation of inhibitor concentration ultimately causes detection problems. This same problem will occur whenever K_3 is large and the inhibitor is a high-molecular weight substance.

D. Experimental Data

There have only been a few published attempts to measure rate constants under isocratic conditions using affinity chromatography.[18,24-28] Typically, measured values of k_{-3} have been 10 to 100-fold lower than expected from solution data.[24-28] We have suggested that these apparent low values were due to attributing too much of the band-broadening to slow kinetics of adsorption and desorption, and too little to the diffusional mass-transfer processes.[18] In particular, the eddy diffusion-mobile phase mass-transfer term has frequently been ignored.[24-26,28]

For example, k_{-3} for NAD applied to immobilized-alcohol dehydrogenase was found to be 100-fold lower than the solution value of 40 sec^{-1} when a 10 μm-diameter silica matrix was used.[26] Repeating our earlier calculations, we would estimate that $H_m = 5 \times 10^{-3}$ cm, $H_{sm} = 9.8 \times 10^{-3}$ cm (assuming $D_m = 5 \times 10^{-6}$ cm^2/sec), and $H_k = 1.6 \times 10^{-3}$ cm. Thus, H_k represents only 9% of the total plate height at $k' = 1$, so it is unlikely that k_{-3} values greater than about 1 sec^{-1} could be determined with any accuracy. On the other hand, the measured plate heights were more than 0.1 cm, or about 10 times greater than the above calculations predict. Thus, either k_{-3} was much smaller than the solution value, or other factors increased the band-broadening, such as poor column packing, excessive extracolumn band-broadening, or nonlinear elution conditions. We would suggest that one should not place much faith in the results of rate-constant measurements unless sufficient supporting work has been done to show that slow adsorption/desorption kinetic band-broadening does indeed dominate over the other sources of band-broadening.

1. Immobilized-Con A Data

An immobilized-Con A system similar to that described in Section III.C. was examined in considerable detail by Muller and Carr[27] and ourselves.[18] Both groups of workers obtained similar raw data, but the conclusions were somewhat different. The results of Anderson and Walters were based on studies of three different immobilized-Con A columns in which the Con A coverage and particle diameters were varied, and two different solutes (PNPM and methylumbelliferyl α-D-mannopyranoside, MUM) were injected.[18]

a. Equilibrium Constants

All three columns exhibited the linear plots of $1/k'$ vs. $[I]$ expected for monovalent solutes.[18] The equilibrium constants were similar on all three columns with values about 2-fold larger than expected from solution studies: K_3 (PNPM) $= 2.4 \times 10^4 M^{-1}$, K_3 (MUM) $= 4.5 \times 10^4 M^{-1}$, K_2 (MDM) $= 8.4 \times 10^3 M^{-1}$. The amounts of Con A in the columns ranged from 16 to 650 nmol.

A very important factor in both the equilibrium- and rate-constant determinations was the use of linear-elution conditions. Figure 5 shows that as the amount of solute injected increased, the peaks became more tailed and retention appeared to decrease. From Figure 5, it appears that linear-elution conditions are approached when the number of moles of sample applied is less than about 1% of the number of moles of active ligand in the column.

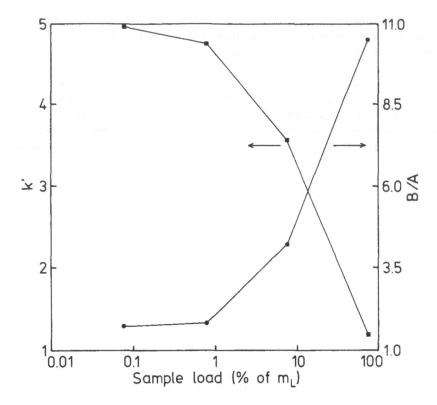

FIGURE 5.   Effect of the sample load (as percentage of total-ligand sites) on the capacity factor (■) and peak asymmetry (●). MUM was applied to a low coverage immobilized-Con A column.

To calculate equilibrium constants it was also necessary to determine m_L, the number of moles of active ligand in the column. A chemical assay tends to overestimate m_L since inactive and/or inaccessible ligands would also be measured. Break-through curves provide a useful measure of the number of active ligand molecules. To perform a break-through analysis a constant concentration of solute is pumped through the column until the adsorption sites are saturated and the concentration of solute leaving the column equals that entering the column (Figure 6). If the curve is symmetric, the break-through volume is the volume at one-half the plateau value. Slow kinetics usually make the curve asymmetric. A computer program can be used to find the point where the areas on either side of the curve are equal (Figure 6). A break-through curve for a nonretained solute should also be run since this volume must be subtracted from the above volume to correct for the column-dead volume. The amount adsorbed will be equal to the concentration of solute applied multiplied by the corrected break-through volume; in some cases this amount will equal m_L. In fact, the measured amount adsorbed represents one point on an adsorption-isotherm plot (Figure 7). If the concentration of solute applied is not large enough, then the equilibrium will not be favorable enough to saturate all of the sites. This is particularly a problem if K_3 is small. Therefore, it is best to make several break-through measurements using various concentrations of solute. The value of m_L is the plateau value of the isotherm plot. However, there are cases where weak nonspecific adsorption of solute takes place in addition to the biospecific adsorption. Even in this situation, it is possible to obtain both K_3 and m_L from the data.[18]

b. Rate Constants

To calculate rate constants from the plate-height data, it was necessary to determine

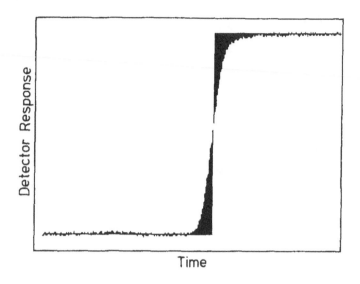

FIGURE 6. Break-through curve for PNPM on an immobilized-Con A column. The darkened regions are the two areas that were matched by the computer in order to find the break-through point.

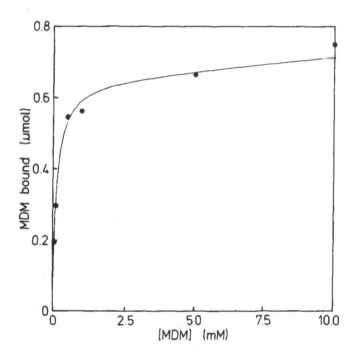

FIGURE 7. Adsorption isotherm plot for MDM on an immobilized-Con A column. Each point was measured using a break-through curve. The values of K_3 and m_L can be calculated from this frontal-analysis data.

several parameters: u, V_m, V_e, V_p, k_{-1}, H_{sm}, and H_m. V_m and u are best determined by injecting a nonretained solute of the same size as the solute of interest. In this case it was assumed that both MUM and PNPM would be able to penetrate all of the pore volume in the column, so pure water was injected and its retention time was taken to be t_m. V_e is determined by injecting a large, totally excluded solute. In this case 0.1 µm-diameter carboxylate microspheres were used, and V_p was calculated by difference.

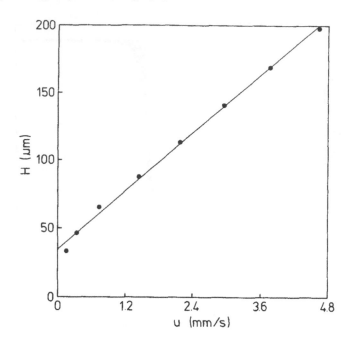

FIGURE 8.   Plate heights for MUM on a blank (diol-bonded silica) column as a function of linear velocity.

In order to calculate H_k from the total plate height experimentally measured, one must independently determine $H_{sm} + H_m$ at each value of k′ and subtract that quantity from H_{tot}. There are several possible approaches for the independent measurement of H_m and H_{sm}. Ideally, one would like to measure $H_m + H_{sm}$ at every k′ used. This approach was tried using PNPM applied to a reversed-phase column.[29] Because of the high-ligand density and nature of the adsorption process, one would not expect any significant kinetic band-broadening. Although this approach allows one to easily vary k′, it was concluded that the results could not be trusted because recent work[8] has shown that a process called "surface diffusion" affects the band-broadening in reversed-phase columns.

Another approach is to prepare plots of H_{tot} vs. u at every k′ value using the affinity column. Since H_m is supposed to be independent of u, such plots would have intercept H_m and slope $(H_{sm} + H_k)/u$.[7] One could determine k_{-1} under conditions where k′ = 0, e.g., by applying the solute to a blank column (no immobilized ligand) or by applying a nonretained analog of the solute to the affinity column. Since $H_k = 0$ when k′ = 0, the slope of the plot of H_{tot} vs. u would be H_{sm}/u, from which k_{-1} could be calculated. H_{sm} could then be recalculated at all of the experimental values of k′. These values of H_{sm} plus the experimental values of H_m could be subtracted from H_{tot} to give H_k; allowing k_{-3} to be calculated. This approach, though tedious, would allow for a possible dependence of H_m on k′.

The approach actually used for the immobilized-Con A data was to measure H_{tot} vs. u only at k′ = 0 using a blank column.[18] Such a plot is shown in Figure 8 and exhibits the linearity expected from the van Deemter Equation with intercept H_m and slope H_{sm}/u. The value of k_{-1} was calculated from the slope and used to recalculate H_{sm} as a function of k′ for the affinity data according to Equation 22. H_m was assumed to remain constant with k′. The values obtained for PNPM using a 5 μm-diameter matrix were $H_m \sim 3 \times 10^{-3}$ cm and $k_{-1} \sim 120$ sec^{-1}. These are in rough agreement with the values predicted earlier. This general approach for determining H_m and H_{sm} is clearly less flexible than the first two methods since it does not allow for H_m to change with k′. Another problem is that the blank

Table 1
DISSOCIATION-RATE
CONSTANTS FOR PNPM ON
THREE IMMOBILIZED-CON
A COLUMNS AS A
FUNCTION OF k'

Matrix	k'	$k_{-3}(sec^{-1})$
5 μm, low	0.1	4.8
coverage	0.3	3.8
	0.7	3.3
	1.3	3.1
	1.9	2.8
	2.3	2.7
10 μm, medium	0.2	2.6
coverage	0.8	1.5
	1.5	1.2
	4.2	0.7
	5.5	0.7
	8.4	0.5
5 μm, high	0.4	3.7
coverage	0.9	3.0
	3.9	1.9
	6.9	1.6
	19.1	1.2
	33.0	0.9

column may not be packed the same as the affinity column or may have a slightly different pore volume. This latter problem can be overcome by measuring H_m and k_{-1} on the affinity column using a nonretained analog of the solute; in this case a good choice would be p-nitrophenyl α-D-galactopyranoside.

Some of the results of the measurements of k_{-3} are given in Table 1. It can be seen that k_{-3} appears to change with k'. Near k' = 1, the 5 μm-particle diameter columns gave k_{-3} ~ 3 sec^{-1}, or about half the literature value of 6 sec^{-1}. This in fact, is quite reasonable since K_3 was twice as large as the literature value; thus k_3, the association-rate constant, was very close to the literature value.

In spite of the reasonable values obtained near k' = 1, the changes in k_{-3} with k' were unexpected and cast doubt on the accuracy of the method. It is our contention that the apparent changes in k_{-3} with k' were really due to errors in calculating H_m and H_{sm} as a function of k', i.e., these terms were underestimated, thereby leading to overestimation of H_k and low values of k_{-3}. Some evidence of this is described below.

First, in Figure 9 is a plot of H_{tot} vs. k' for all three columns. If H_k was the dominant term, then there would be a maximum at k' = 1 and a sharp dropoff at higher k'. In fact for the 10 μm matrix no maximum was seen, while for the 5 μm matrices there was a maximum but not much of a decrease at higher k'. This suggests that $H_{sm} + H_m > H_k$ for the 10 μm matrix and that $H_{sm} + H_m \sim H_k$ for the 5 μm matrices. If this is true, then the values of H_{sm} and H_m calculated as k' increased must have been low.

Some additional evidence supporting this is the observation that the peak asymmetries for all three columns did not change appreciably with k'. As pointed out earlier, this indicates that $H_{sm} > H_k$.

An important point is that H_m and k_{-1} were determined at k' = 0. Any errors in these measurements, and in V_m, V_p, and V_e, would tend to be amplified when H_m and H_{sm} were calculated at higher k' values. For this reason and because H_k becomes smaller relative to

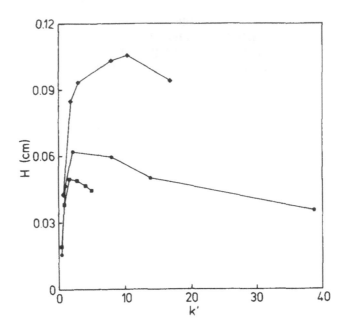

FIGURE 9. Experimental plots of plate height vs. capacity factor for MUM applied to a low coverage, 5 μm-diameter matrix (■); a medium coverage, 10 μm-diameter matrix (♦); and a high coverage, 5 μm-diameter matrix (●).

$H_m + H_{sm}$ at $k' > 1$ one would expect the values of k_{-3} measured at high k' to be less accurate.

If the problem were simply errors in calculating $H_m + H_{sm}$ as a function of k', then fitting the H_{tot} vs. k' data to the equation for H_m, H_{sm}, and H_k should provide an alternative method for determining k_{-3}. This approach yielded reasonable values for k_{-3} but the fits were not good.[18] This suggests a more fundamental discrepancy between the theory and the experimental data. This could be due to nonideal behavior, such as nonuniform packing of the silica, or to deficiencies in the theory, e.g., the unknown dependence of H_m on k'. These factors suggest that to make accurate measurements of H_k, H_k must be even bigger relative to $H_{sm} + H_m$ than was suggested by the example calculations. It appears that for a small solute like PNPM it is difficult to measure k_{-3} values greater than 1 sec^{-1} even with 5 μm particles, while for macromolecular solutes the upper limit would be two orders of magnitude smaller.

2. Future Prospects for Isocratic Measurements

From the discussion above it is clear that more work needs to be done to clarify how well the chromatographic theory predicts the behavior of real columns, especially as a function of capacity factor. Even if the theory is not entirely accurate one should still be able to measure k_{-3} if the H_m and H_{sm} contributions to the plate height can be diminished. Recent developments in the production of small porous, nonporous, and pellicular matrices suggest that future technology will allow us to make more accurate measurements and to measure faster dissociation-rate constants.

Wade et al.[35] have recently described a method for the measurement of rate constants under nonlinear elution conditions. This important advance may significantly simplify and broaden the applicability of isocratic rate constant measurements since larger sample loads may be applied.

IV. SPLIT-PEAK METHOD

The fundamental problem of isocratic studies is the need to know the theoretical dependence of each plate-height term on k'. This problem has led to the search for other chromatographic means for measuring either k_3 or k_{-3} which do not require detailed knowledge
of how each term depends on k'. In this chapter, the "split-peak" method for determining
association-rate constants is described.

A. Theory

Imagine the steps leading to adsorption of a solute molecule in a chromatographic column.
First, the molecule must leave the flowing-mobile phase and enter the stagnant-mobile phase
of the pores. Then, it must diffuse through the liquid to the surface, where it may finally
adsorb. It is not difficult to imagine that at high-flow rates the molecule might not have
time to complete these steps and would instead pass through the column unretarded and
elute in the void or excluded volume. This is the basis of the split-peak method.

Giddings and Eyring developed some of the theory necessary for the split-peak measurement of rate constants.[30] More recently, the theory was extended to include a diffusional-
mass transfer step but under conditions of irreversible adsorption.[22]

$$E_e \underset{k_{-1}}{\overset{k_1}{\rightleftharpoons}} E_p \tag{31}$$

$$E_p + L \xrightarrow{k_3} E_p - L \tag{32}$$

The rate constants and other symbols used here are the same as in the isocratic theory.

Assuming linear-elution conditions, the result is[22]

$$\frac{-1}{\ln f} = F\left(\frac{1}{k_1 V_e} + \frac{1}{k_3 m_L}\right)$$

$$= \frac{u_e}{L}\left(\frac{1}{k_1} + \frac{V_e}{V_p k_3^*}\right) \tag{33}$$

The fraction of the solute which elutes in the column void or excluded volume is "f". The
fraction $b = 1 - f$ remains irreversibly adsorbed inside the column.

Just as in the isocratic theory derived by Hethcote and DeLisi,[11-14] Equation 33 only
accounts for the stagnant-mobile phase and slow-adsorption band-broadening processes, but
not for any interparticle band-broadening processes.

Examination of Equation 33 reveals several important facts. First, if the term on the right
side of the equation is large, then a significant fraction of the solute will elute in the void
volume of the column. The right side of the equation can be made large by increasing the
flow rate, decreasing the column length (decreased V_e and m_L), decreasing the ligand density
(decreased m_L), or by increasing the particle size (decreased (k_1)) or any other factors which
would decrease k_1 or k_3.

The rate of adsorption will be limited by the stagnant-mobile phase mass-transfer rate if
$1/k_1 V_e > 1/k_3 m_L$. This will be called a "diffusion-limited" system. If $1/k_3 m_L > 1/k_1 V_e$,
then the system will be called "adsorption-limited" and it should be possible to measure
k_3.

Normally, chromatography is done under conditions where $f \sim 0$. It is apparent that more unusual conditions, i.e., shorter columns and higher flowrates, will be necessary for split-peak studies. The earlier discussion indicated that isocratic measurements should be made using conditions where $N > 50$, so one can anticipate $N \ll 50$ for split-peak experiments.

The major advantage of the split-peak method is that only peak areas are measured. This means that injector, detector, and other sources of extra-column band-broadening will not affect the results. This is not only convenient, but should improve the precision of the results since peak areas can be measured much more precisely than peak variances.[17] In fact, the nonretained peak can simply be collected in a test tube and assayed later.

Another advantage of the split-peak method is that no competing inhibitor is needed. Although the bound peak should be removed after each run, it can be eluted by nonspecific (e.g., a pH change) as well as specific means. This makes the split-peak method particularly well-suited for antibody-antigen and other macromolecule-macromolecule complexes.

Equation 33 is best applied by making a series of measurements at several flow rates and plotting $-1/\ln f$ vs. F. This will be called a "split-peak plot". A straight line with an intercept of zero should be observed; the slope will equal $1/k_1V_e + 1/k_3m_L$. It is preferable to utilize conditions such that $f \geq 0.1$. This will minimize any errors due to impurities in the sample. These impurities might adsorb more quickly or slowly than the major solute and could cause curvature of the split-peak plots.

Just as in the isocratic method, independent estimates of k_1, V_e, and m_L are needed to calculate k_3. However, there is no k' dependence of Equation 33, so subtraction of the $1/k_1V_e$ contribution should be more reliable than subtraction of H_{sm} in the isocratic method.

Another interesting difference between the isocratic and split-peak methods is the effect of heterogeneity on the measured-rate constant. For example, assume 50% of the ligands have $k_{-3} = 1.0$ and 50% have $k_{-3} = 10.0$. It is simple to show that the apparent-rate constant measured using the isocratic plate-height method will be

$$1/(0.5/1.0 + 0.5/10.0) = 1.8$$

For the split-peak method, if k_3 has the same values, the result will be

$$0.5 \times 1.0 + 0.5 \times 10.0 = 5.5$$

Thus, the isocratic method tends to weight the apparent-rate constant toward the lower end of the range, while the split-peak method weights it more toward the higher end of the range.

B. Computer Simulations

The problem with the split-peak theory is that it assumes irreversible adsorption, i.e., $k' = \infty$, $k_{-3} = 0$. If this were really the case, one would probably not be interested in doing the chromatography in the first place. To examine the more practical cases in which k' has a finite value, the two-step reversible-kinetics simulation program described earlier has been utilized. This program is also useful in that it can describe the transition from "split-peak" to "normal" chromatographic behavior.

Figure 10 shows a computer simulation of a kinetically-limited case in which $k' = 5$, i.e., the center of the peak should be at 6 void volumes. For the longest column, 29-plates are generated and the peak is seen to be nearly symmetric. As the column is shortened 4-fold and 7.2 plates are generated, the peak becomes broader (relative to the void volume, which is also decreasing) and more tailed. As the length decreases an additional 2.5-fold and the plate number decreases to 2.9, some solute begins to elute at the void volume and a double peak is observed. A further 4-fold decrease in the length and N yields a peak which mostly elutes at the void volume, with the retained material eluting as a long tail.

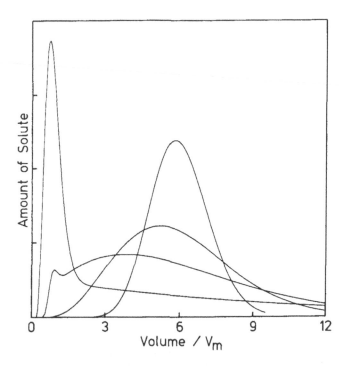

FIGURE 10. Computer simulation of the transition from a "normal" peak using a long column (or low flow rate) to a "split" peak using a short column (or high flow rate) for a solute with $k' = 5$. Conditions: L = 0.05, 0.2, 0.5, and 2.0 cm (from most-skewed to most-symmetric), u_e = 0.05 cm/sec, k_1 = 1000 sec^{-1}, k_3^* = 1.0 sec^{-1}, and $V_p/V_e = 1$.

When the adsorption is reversible, it is clear that the "bound" and "free" fractions are really one continuous peak. If k' is fairly large, the chromatogram will appear as a sharp peak at the void volume followed by a very long, low-concentration tail, i.e., the bound fraction will slowly bleed from the column. How small must k' be so that this bleeding will be negligible? Figure 11 shows part of a computer simulation performed at $k' = 3$, 10, and ∞. When $k' = \infty$, the fractional area under the nonretained peak is the value which would be calculated from Equation 33, in this case 0.50. When $k' = 10$, the retained fraction elutes as a very long, low tail. The free fraction can still be estimated with acceptable accuracy by somewhat arbitrarily choosing the "end" of the void peak. When $k' = 3$, the bound fraction bleeds off too rapidly and would prevent the accurate measurement of f. Thus, it appears that the split-peak method may be applied when $k' \geq 10$. This is likely to be true in the majority of the cases of interest.

C. Example Calculations

In this section, calculations will be performed to indicate the applicability of the split-peak method.

Since both V_e and m_L depend on column length, it is more convenient to rewrite Equation 33 in the following form

$$\frac{-1}{\ln f} = \frac{F}{V_m} \left(\frac{V_m}{k_1 V_e} + \frac{V_m}{k_3 m_L} \right) \tag{34}$$

Examination of Equation 34 indicates that there are three experimental parameters that one can adjust: F, V_m, and m_L/V_m. By increasing m_L/V_m, i.e., immobilizing a higher density of

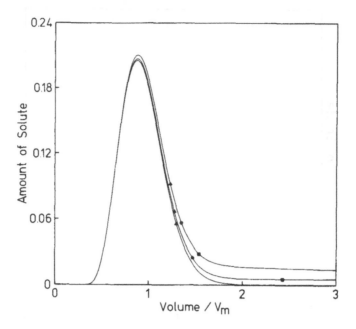

FIGURE 11.   Computer simulation of the nonretained and early part of the retained peak for split-peak data when $k' = 3$ (top curve), $k' = 10$ (middle), and $k' = \infty$ (bottom). The theoretical value of "f" is 0.50. If the area of the nonretained peak was measured up to the point marked, the error in $-1/\ln f$ would be -10% (▲), 0% (●) and $+10\%$ (■).

active ligands on the support surface, one would tend to shift the system toward the diffusion-limited extreme. A low density of ligands would thus be useful for ensuring adsorption-limited conditions.

Of course, the choice of solute and matrix will have a large impact on whether the system is diffusion- or adsorption-limited. Since k_1, like k_{-1}, is proportional to D_m and inversely proportional to d_p^2 (see Equations 9 and 23), the diffusion-limited case will be less likely when the solute and matrix diameter are both small, just as in the isocratic method.

Once the term in parenthesis in Equation 34 is fixed, then F and V_m can be adjusted to provide the desired free fraction (e.g., $f \sim 0.2$). If the term in parenthesis is too small, then F must be very large or V_m very small. This poses practical problems of column stability at high-flow rates or detection and overloading problems when V_m and m_L are small.

1. DNP-Antibody Kinetics

Consider first the possibility of measuring k_3 for the immobilized antibody-DNP hapten system discussed in the previous section. The basic data, assuming $V_e = V_p$ and $d_p = 5$ μm are as follows

$$k_{-3} = 500 \text{ sec}^{-1}$$

$$k_3 = 5 \times 10^8 \, M^{-1} \text{ sec}^{-1}$$

$$k_{-1} = k_1 = 240 \text{ sec}^{-1}$$

One might anticipate difficulty with this system since k_3 is probably close to the maximum-rate constant as limited by diffusion.[20] Since diffusion distances are inherently longer when one of the species is immobilized on a surface than in a homogeneous solution, this system should be diffusion-limited.

To try to make the system adsorption-limited, one can adjust m_L/V_m so that the kinetic part of the slope term in Equation 34 will be 10-fold greater than the diffusional part of the slope, $V_m/k_3m_L \geq 10 \, V_m/k_1V_e$. Using the data above, $m_L/V_m \leq 2.4 \times 10^{-8} \, M$. Assuming $f = 0.20$, from Equation 34, $F/V_m = 6.8 \, sec^{-1}$. If a flow rate of 1 mℓ/min is chosen, then $V_m = 2.5 \, \mu\ell$. This is a very tiny column and would contain only 6×10^{-14} mol of ligand!

Although this would appear to present formidable technical problems there is an even more important fundamental problem. Use of a very low-ligand density support decreases k'. From Equation 15 with $[I] = 0$, one can calculate that $k' = 0.024$ under the above conditions. This is much lower than the minimum value of 10 specified earlier.

This problem can be examined more directly by noting that $k' = m_{EL}/m_E$ and $K_3 = (m_{EL}/A)/((m_E/V_m)(m_L/A))$ so that

$$k' = \frac{K_3m_L}{V_m} = \frac{k_3m_L}{k_{-3}V_m} \tag{35}$$

Then Equation 34 can be rewritten as

$$\frac{-1}{\ln f} = \frac{F}{V_m}\left(\frac{V_m}{k_1V_e} + \frac{1}{k_{-3}k'}\right) \tag{36}$$

Thus, if a minimum value of k' is set, then there is a maximum value for the kinetic part of the slope. For the DNP case, $1/k_{-3}k' = 2 \times 10^{-4} \, sec$ (with $k' = 10$); since $V_m/k_1V_e = 8.3 \times 10^{-3} \, sec$, it is impossible to make split-peak measurements of k_3 in this case.

2. Con A — Sugar Kinetics

From the previous PNPM-Con A calculations

$$k_{-3} = 6.2 \, sec^{-1}$$

$$k_3 = 5.4 \times 10^4 \, M^{-1} \, sec^{-1}$$

$$k_{-1} = k_1 = 240 \, sec^{-1}$$

Using Equation 36, $V_m/k_1V_e = 8.3 \times 10^{-3} \, sec$ and $1/k_{-3}k' = 0.016 \, sec$ when $k' = 10$. Thus, at best, both terms will contribute about equally to the slope of the split-peak plot. This would make it difficult to obtain an accurate value of k_3. Furthermore, from Equation 35, $m_L/V_m = 1.2 \times 10^{-3} \, M$, and from Equation 34 or 36, $F/V_m = 26 \, sec^{-1}$ when $f = 0.2$. If $F = 1 \, m\ell/min$, $V_m = 0.6 \, \mu\ell$. Once again, it appears that the split-peak method could not be used because the column would have to be exceedingly small.

3. Protein A — IgG Kinetics

From the earlier estimates for this system

$$k_{-3} = 10^{-3} \, sec^{-1}$$

$$k_3 = 4 \times 10^5 \, M^{-1} \, sec^{-1}$$

$$k_{-1} = k_1 = 2.4 \, sec^{-1}$$

From Equation 36, $V_m/k_1V_e = 0.83 \, sec$ and $1/k_{-3}k' = 100 \, sec$ when $k' = 10$. Here is finally a situation in which adsorption is much slower than diffusion. If $k' = 10$, then $m_L/V_m = 2.5 \times 10^{-8} \, M$. This is a rather small ligand density, so instead let k' and

m_L/V_m increase so that the system just meets the criterion for kinetic control, i.e., $1/k_{-3}k'$ $\geq 10V_m/k_1V_e$; then $k' = 120$ and $m_L/V_m = 3 \times 10^{-7} M$. From Equation 36, $F/V_m = 0.068 \text{ sec}^{-1}$ for $f = 0.2$; if the flow rate is chosen to be 1 mℓ/min, $V_m = 0.25$ mℓ. This would correspond to a column 1.8 × 0.46 cm containing 0.075 nmol active ligand. Thus, the main problem with this system would appear to be detection or overloading since m_L is so small.

These example calculations show that the split-peak method is most applicable to systems in which K_3 is large and k_3 moderately small. Many antibody-antigen complexes should have these characteristics.

D. Experimental Data
1. Protein A — Immunoglobulin G
In the experimental work the immobilized-protein A — IgG system discussed above was used.[22] The major difference between the calculations and the experiment was that m_L/V_m was about 100-fold higher than calculated. According to the calculations, this would make the system diffusion-limited. As will be seen, this was the case for some of the columns tested. The remaining columns had either larger values of k_1 or smaller values of k_3 and proved to be adsorption-limited.

The protein A-IgG system was chosen because of the observation that IgG applied to a 5 cm-long column at 1 mℓ/min was totally adsorbed but if the column length was 0.64 cm some of the IgG eluted in the void volume. This behavior could be blamed on either true-kinetic behavior or on overloading with IgG. The literature of affinity chromatography contains many examples of chromatograms in which some of the active solute has eluted in the void volume while the remainder was tightly adsorbed. In most cases, we would speculate that this was due to applying too much sample rather than due to the split-peak effect (e.g., see References 31 and 32). This was not true for the immobilized-protein A, however, since split-peaks were observed at all sample sizes.

Since this behavior had not been observed for other biochemical systems in similar small columns[33] it was suspected that the slow kinetic behavior might be a result of the immobilization process. To examine this further three different methods were used to immobilize protein A on diol-bonded silica. Details of the methods can be found elsewhere;[22] they will simply be referred to here as the CDI, EA, and SB methods.

The columns were first examined by break-through (frontal) analysis with IgG to determine m_L. Next, the free and bound peak areas were measured at various flow rates. As the flow rate increased, the size of the void peak increased and the retained peak (which was eluted with a pH change) decreased (Figure 12). Plots of $-1/\ln f$ vs. F were made and exhibited the expected linear behavior (Figure 13). The plots were made at several sample sizes to ensure linear-elution conditions, however, some change in the slope with sample size was observed (Figure 14). This will be discussed more below but for now simply note that the slopes were extrapolated to zero sample size to obtain the values of the slopes used in the calculation of rate constants.

The portion of the slope due to slow diffusion, $1/k_1V_e$, was determined in two ways. First, IgG was isocratically eluted from a blank column (no protein A) and the plate-height data as a function of linear velocity (see Figure 8) was treated as in the isocratic section. The measured value of k_1 was 4.0 sec^{-1}. Second, a split-peak experiment was performed using a reversed-phase support.prepared from the same silica used in the affinity experiments. From Equation 33, one would expect that the high-ligand density and presumably fast-adsorption kinetics would make the reversed-phase column diffusion-limited. Assuming this to be the case, k_1 was determined to be 8.6 sec^{-1} from the slope of split-peak plots. The factor of two difference in the two methods was probably due to heterogeneity as discussed in the theory section. V_e was measured using 0.1 μm carboxylate microspheres.

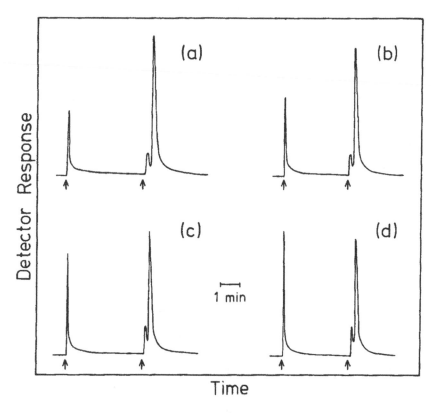

FIGURE 12. Chromatograms of IgG applied to 6.4 × 4.1 mm columns of immobilized-protein A showing the increasing split-peak behavior at flow rates of 1.5 (a), 1.75 (b), 2.0 (c), and 2.25 (d) mℓ/min. The IgG was applied at the first arrow. The bound IgG was eluted by a pH change at the second arrow.

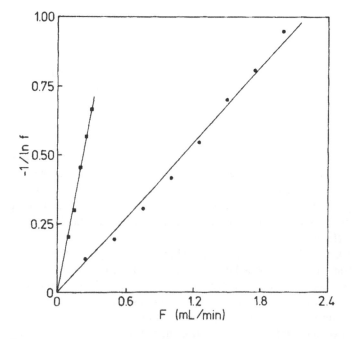

FIGURE 13. Split-peak plots of IgG applied to protein A immobilized using the CDI (■) and SB (●) methods.

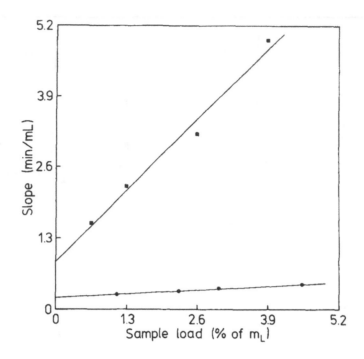

FIGURE 14.   Extrapolation of the split-peak plot slopes to zero-sample size. The symbols are the same as in Figure 13.

Table 2
SPLIT-PEAK KINETIC DATA FOR IgG
APPLIED TO VARIOUS IMMOBILIZED-
PROTEIN A COLUMNS

Column	m_L(nmol)	Split-peak slope (sec/mℓ)	$1/k_1 V_e$(sec/mℓ)	$k_3(M^{-1}\ sec^{-1})$
CDI-500[a]	1.4	52.0	4—9.	1.5—1.7 × 10^4
EA-500	1.6	11.0	4—9.	9—30. × 10^4
SB-500	3.2	13.0	4—9.	4—8. × 10^4
SB-50	1.0	8.7	0.3—1.3	1.2—1.3 × 10^5

[a]   Pore size of the matrix in Å.

Comparison of the independently measured values of $1/k_1 V_e$ with the measured split-peak slopes from the affinity columns (Table 2) indicates that only one of the 500-Å pore-size columns was primarily adsorption-rate limited. A value of $1.6 \times 10^4\ M^{-1}\ sec^{-1}$ was calculated for protein A immobilized using the CDI method. Adsorption appeared to be faster on the other two columns and so k_3 could not be accurately calculated since the columns were diffusion-limited.

To determine k_3 for SB- or EA-immobilized protein A required the use of either a more diffusionally efficient matrix or a lower-ligand concentration. The latter approach would have presented severe problems of detection, so a more efficient matrix was found. A 50-Å pore-size matrix was chosen because the pores were smaller than IgG; thus it should have behaved like a nonporous bead. In fact, it behaved more like a pellicular matrix because it was not only chromatographically more efficient but also appeared to have a reasonably large surface area.

With this matrix, the split-peak slope using the SB-immobilization method decreased slightly to 8.7 sec/mℓ. The values of k_1 determined by isocratic (36 sec^{-1}) and split-peak (149 sec^{-1}) methods increased. Overall, the data in Table 2 shows that this column was now adsorption-rate limited with $k_3 = 1.2 \times 10^5 \, M^{-1}sec^{-1}$.

Based on this data it was speculated that the SB-immobilization method gave immobilized-protein A, whose adsorption-rate constant of $1.2 \times 10^5 \, M^{-1}sec^{-1}$ was about the same as for free-protein A in solution (however, no literature data was available to support this). It was also thought that the CDI-immobilization method altered the kinetics of the protein A, perhaps by distorting the molecule or hindering access to the binding site, and decreased k_3 by nearly 10-fold to $1.6 \times 10^4 \, M^{-1}sec^{-1}$.

2. Effects of Heterogeneity

Although this application of the split-peak theory appeared to be successful, some facts were puzzling: the change of the split-peak slope with sample size and the differences in k_1 measured by the isocratic and split-peak methods. It is now believed that these problems can be traced to heterogeneity in k_1 and/or k_3 values.

With the CDI-immobilization method, it is not likely that all of the protein A molecules had their adsorption-rate constants decreased exactly the same 10-fold amount, but rather that some were not reduced at all while others were reduced more than 10-fold. If this were the case, there would be many subpopulations of protein A molecules, each with its own m_L and k_3. When a small amount of IgG was applied to such a column, most of the adsorption would take place on the fastest adsorption sites, and the split-peak slope would be small. As the sample size increased, the fast sites would soon be overloaded and more adsorption would take place on the slower sites. This would cause an increase in the split-peak slope. This is exactly the trend observed in Figure 14. Furthermore, the most heterogeneous material should be most sensitive to sample size; this too, is supported by the slopes of the CDI- and SB-immobilized protein A plots in Figure 14.

Therefore, if heterogeneity is present, one should regard the calculated-rate constants as "apparent" values since they are some weighted average of the actual range of rate constants. Also, the extent of kinetic heterogeneity can be qualitatively judged by the steepness of plots like Figure 14. This is potentially useful information which would be difficult to obtain by other means.

These observations are further supported by the isocratic and split-peak values of k_1 measured using the same matrix. As pointed out earlier, one would expect the isocratic data to be weighted toward the slowest k_1 values and the split-peak data to be weighted toward the highest k_1 values, as was observed experimentally.

3. Future Prospects for Split-Peak Measurements

Just as with the isocratic method, the range of applicability of the split-peak method will increase as new, more efficient matrices are developed. The split-peak method should prove valuable for more than just the measurement of rate constants. It provides a new way to optimize column performance (e.g., choosing the flow rate where $f \leq 0.01$) and to kinetically compare columns which might appear to be identical in terms of total-adsorption capacity (e.g., all four columns in Table 2 have similar value of m_L). It also provides a simple way to assess the heterogeneity of affinity columns.

V. PEAK-DECAY METHOD

The peak-decay method is a new method which was developed specifically for the measurement of dissociation-rate constants. The basis for this method can be thought of as follows. Imagine a very short affinity chromatographic column which is initially saturated with solute, followed by removal of the excess-free solute. Next, while buffer flows through the column, conditions are changed so that the readsorption rate (k_3) is zero, while the dissociation rate (k_{-3}) is unaffected. The result will be that whenever a solute molecule dissociates from a ligand molecule, the solute molecule will not readsorb but will immediately elute from the column. It is apparent that the peak elution profile will be a first-order exponential decay with rate constant k_{-3}.

The key to this method is to prevent readsorption during the second part of the experiment, i.e., make $k' \sim 0$. Although this could be done by a pH change or other denaturing conditions, k_{-3} might also change. Therefore, as in the isocratic method, one must use a competing inhibitor in the mobile phase to decrease retention. By occupying any free-ligand sites, the inhibitor prevents the solute from readsorbing on these sites.

This method should be applicable to either "normal" or "reversed-role" affinity chromatography. Thus, the inhibitor could be an analog of either the affinity ligand or the solute. In the case of a macromolecule-macromolecule complex, one can utilize a labeled protein as the solute and the same protein, but not labeled, as the inhibitor.

A. Theory

It is assumed that after free-excess solute is washed from the column, all of the remaining solute is bound to ligand molecules. At time zero (when the competing inhibitor is applied to the column), the E-L complex begins to irreversibly dissociate. It is also assumed that the column is of length approaching zero so that once the free solute diffuses out of the stagnant mobile phase, it is immediately washed out of the column. The kinetics for this situation can then be written as a two-step irreversible reaction

$$E_p - L \xrightarrow{k_{-3}} E_p + L \tag{37}$$

$$E_p \xrightarrow{k_{-1}} E_e \tag{38}$$

The symbols again are the same as those used previously.

If m_{E_p} and m_{E_e} are initially assumed to be zero, and if $k_{-3} < k_{-1}$, the elution profile of the peak is given by [34]

$$\frac{dm_{E_e}}{dt} = \frac{m_{E_0} k_{-1} k_{-3}}{k_{-1} - k_{-3}} (e^{-k_{-3}t} - e^{-k_{-1}t}) \tag{39}$$

Where m_{E_0} is the initial number of moles of solute adsorbed.

The natural logarithm of the profile is

$$\ln \frac{dm_{E_e}}{dt} = \ln\left(\frac{m_{E_0} k_{-1} k_{-3}}{k_{-1} - k_{-3}}\right) + \ln(e^{-k_{-3}t} - e^{-k_{-1}t}) \tag{40}$$

In general, the decay will not be linear. If $k_{-3} << k_{-1}$, then

$$\ln \frac{dm_{E_e}}{dt} = \ln(m_{E_0} k_{-3}) - k_{-3}t \tag{41}$$

In this case, the slope of the logarithm of the peak decay profile will be $-k_{-3}$.

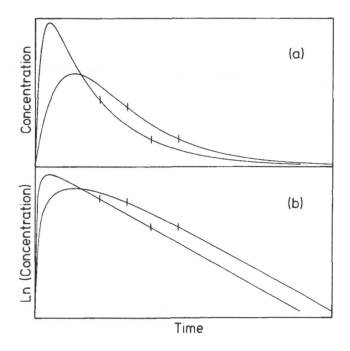

FIGURE 15. Peak-decay profiles (a) and the natural logarithm of the profiles (b) calculated from Equation 39 for the conditions $k_{-1}/k_{-3} = 10$ (taller peak) and $k_{-1}/k_{-3} = 2$ (shorter peak). The vertical marks indicate the region over which the slope was measured.

B. Computer Simulations

While Equations 39 to 41 are useful for describing the expected peak profiles under ideal conditions, they do not indicate how these conditions can be approximated in practice. For example, it is unreasonable to expect that $k_{-1} = 0$ since k_1 and k_{-1} are of similar magnitude.

To establish conditions under which Equation 41 will be obeyed, the computer program described earlier which models the two-step reversible kinetic system was used. One would expect the reversible-kinetic situation (Equations 5 and 6) to reduce to the irreversible case (Equations 37 and 38) when k' is small and the residence time of a nonretained solute in the column (t_m) is small compared to $1/k_{-3}$ and $1/k_{-1}$.

In the computer model, the solute is initially assumed to be uniformly and irreversibly adsorbed along the length of the column. A plug of competing inhibitor begins at time zero to travel down the column with velocity u_e. Behind the competing inhibitor front, k' drops to a constant predetermined value and the solute begins to undergo dissociation, readsorption, and diffusion just as in the isocratic method.

The simulation does not take into account broadening and dilution of the competing inhibitor front, nor does it take into account the finite number of ligand sites (as before, the model assumes linear-elution conditions). Inclusion of the former process in the model would probably cause the elution profile to be broader, while the latter would cause it to be narrower.

Figure 15a shows some of the peak-decay profiles calculated from Equation 39 for various values of k_{-1}/k_{-3}. Identical values are obtained from the computer-simulation model if k' is very small and the column length is short. When $k_{-3} << k_{-1}$ the peak onset is sharp with a long tail. When $k_{-3} \sim k_{-1}$ a more normal-looking peak is observed. From the logarithm plot (Figure 15b), the curvature predicted by Equation 40 when $k_{-1} \sim k_{-3}$ is apparent. In such a situation, an arbitrary choice of the portion of the profile to use in determining the slope must be made. All of the slopes from computer simulations were

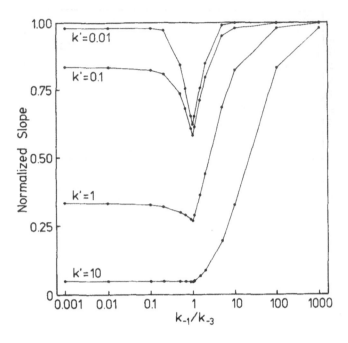

FIGURE 16. Plot of the normalized slope from computer simulations of peak-decay profiles as a function of k_{-1}/k_{-3} and k' for the conditions $V_p = V_e$ and $L_{col} = 0$.

measured in the following way: the times corresponding to the top of the peak profile and the point where 99% of the solute had eluted were determined, the time between these points was divided into five equal segments, and the slope of the second segment (from the peak) was determined by linear least-squares fitting of the points. The segment used is marked in Figure 15. This particular segment was chosen because it most closely duplicated the region which was used in the experimental data.

In cases where the peak profiles are curved, the measured slopes are always less than the smaller of k_{-1} or k_{-3}. Therefore, more accurate values for the slopes are obtained if the slope is measured further out along the tail of the peak. Experimentally, however, detector noise and uncertainty in the position of the baseline prevent the use of data in this region (see, e.g., Figure 19).

Utilizing the computer-simulation model it was possible to determine the slopes of peak-decay profiles (and hence the apparent-rate constants) over a wide range of k_{-1}/k_{-3} and k' values. In general, the logarithm of the profile was curved when $k_{-1} \sim k_{-3}$, but even in this situation became nearly linear at high k'.

Figure 16 is a plot of the normalized slope, i.e., the slope of the "second segment" described above divided by the smaller of k_{-1} or k_{-3}. This normalized slope never exceeds 1, so just as in the isocratic method, if diffusional effects are not negligible, low values for k_{-3} will be obtained.

The most notable feature of Figure 16 is the sharp minimum at $k_{-1}/k_{-3} = 1$. Of course this occurs because both k_{-1} and k_{-3} are contributing to the decay (see Equation 39) and the profile is broader and has a shallower slope than if only one of the rate constants was limiting. As a result, k_{-1} can only be measured if $k_{-1} \leqslant k_{-3}/10$ and k_{-3} can only be measured if $k_{-3} \leqslant k_{-1}/10$ (see Figure 16).

Figure 16 shows that as k' increases the accuracy of the apparent-rate constants determined will decline; this is because multiple adsorption and desorption steps begin to take place and cause additional band-broadening. Interestingly, at high k' the plot is very asymmetric

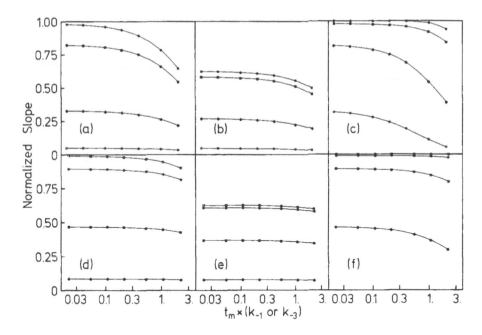

FIGURE 17. Plots of normalized slope from computer simulations of peak-decay profiles as a function of void time multiplied by the smaller of k_{-1} or k_{-3}. $V_p/V_e = 1$ in (a to c), $V_p/V_e = 0.1$ in (d to f). In (a) and (d), $k_{-1}/k_{-3} = 0.1$. In (b) and (e), $k_{-1} = k_{-3}$. In (c) and (f), $k_{-1}/k_{-3} = 10$. $k' = 0.001$ ($\square$), 0.01 ($\bullet$), 0.1 ($\blacksquare$), 1.0 ($\blacklozenge$), and 10.0 ($\bigcirc$).

and indicates that when k_{-1}/k_{-3} is large k_{-3} can still be determined accurately at high k', but when k_{-1}/k_{-3} is small k_{-1} cannot be determined accurately. This difference is due to the fact that in the former case, as soon as the adsorbed complex dissociates, the solute immediately diffuses out of the pore and is washed from the column. In the latter case, adsorbed solute and free solute in the pore volume are essentially in equilibrium and so the solute diffuses out of the pore more slowly than if all the solute were free.

When the capacity factor is less than about 0.01, then the measured slopes begin to approach the true-rate constants and the peak-decay profiles begin to agree with those generated from Equation 39. Equation 39 shows no dependence of the profiles on V_p/V_e. This was also observed in the computer simulations, but only when k' was very small. For example, the figure analogous to Figure 16 but with $V_p/V_e = 0.1$ looked almost the same but with the capacity factors shifted 10-fold, i.e., what is labeled $k' = 10$ in Figure 16 would be $k' = 1$ in the new figure.

The discussion above was limited to the case of zero-column length. As the length increases, more multiple adsorption and desorption steps will take place, and the peak-decay profiles will become broader and less steep. This is illustrated in Figure 17, in which the normalized slope is plotted vs. the column-void time multiplied by the smaller of k_{-1} or k_{-3}. The data is also presented as a function of k' and V_p/V_e. The limiting value of the slope as t_m approaches zero corresponds to the data in Figure 16.

Figure 18 shows two of the computer-simulated peak-decay profiles used in preparing Figure 17c. As the void time and/or capacity factor increase, the decay profiles become broader. The logarithm of the profile usually becomes curved and not as steep, resulting in low apparent values of the rate constants. Figure 17 indicates that as the column length decreases or the flow rate increases (both of which decrease t_m), a limiting value of the slope is observed. The point where this plateau value is reached is a function not only of t_m but also of V_p/V_e, k', and k_{-1}/k_{-3}. For example, it is apparent that measurements of k_{-3}

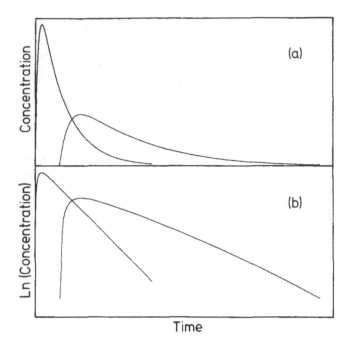

FIGURE 18.   Computer-simulated peak decay profiles (a) and the natural logarithms of the profiles (b) when $k_{-1}/k_{-3} = 10$ and $V_p/V_e = 1$. For the taller peak, $t_m k_{-3} V_p/V_m = 0.01$ and $k'V_m/V_p = 0.02$. For the shorter peak, the quantities are 1.0 and 2.0, respectively.

(Figure 17c) will be less sensitive to changes in both column length and k' than measurements of k_{-1} (Figure 17a).

Although Figures 16 and 17 may be a bit confusing, a careful examination reveals that the data can be reduced to a set of three criteria which will allow accurate (error $\leq$ 5%) determination of k_{-1} or k_{-3}.

To accurately measure k_{-1}

$$k_{-1}/k_{-3} \leq 0.1 \tag{42}$$

$$t_m k_{-1} V_p/V_m \leq 0.05 \tag{43}$$

$$k'V_m/V_p \leq 0.02 \tag{44}$$

To accurately measure k_{-3}:

$$k_{-1}/k_{-3} \geq 10 \tag{45}$$

$$t_m k_{-3} V_p/V_m \leq 0.2 \tag{46}$$

$$k'V_m/V_p \leq 0.2 \tag{47}$$

Curved logarithmic plots of the data occur, as pointed out above, when k_{-1} and k_{-3} are too close in size or when t_m is too large. One might wonder whether a linear-logarithmic plot would itself guarantee that the criteria above are met. The data indicate that there are many cases where k' is large and the logarithmic plots are linear, but the apparent slopes are much smaller than the true-rate constants.

1. Effect of Heterogeneity

If there is heterogeneity in the values of k_{-1} or k_{-3}, then Equation 39 could be expanded to include more terms summed up, each with its own m_{E_o}, k_{-1}, and k_{-3}. The logarithmic plots would then be the sum of several exponential terms, leading to a plot which flattens as time increases (just the opposite of Figure 15b, where the plot becomes steeper with time).

If the slope is measured late in the tail of the peak, it always approaches the smallest of the individual-rate constants because the faster-decaying solute molecules have already left the column. If the slope is measured as described earlier, the apparent-rate constant turns out to be quite close to the apparent value that would have been obtained from an isocratic experiment, i.e., weighted toward the smallest-rate constant.

There are, however, other major differences between the isocratic and peak-decay methods. A major advantage of the peak-decay method, like the split-peak method, is that the calculations do not depend directly on k'. Another advantage is that the slope of the logarithm of the peak-decay profile can be determined with excellent precision compared to peak variances. A final advantage is that since dissociation only needs to take place once (rather than many adsorption/desorption steps in an isocratic run), one can determine even very small k_{-3} values in reasonable periods of time.

One might anticipate one other advantage. Since the column can be saturated with solute because linear-elution conditions are unnecessary, detection ability should improve. However, the necessity of introducing a high concentration of inhibitor into the column at time zero causes refractive index changes which may severely interfere with detection.

C. Example Calculations

As before, the example calculations will demonstrate the range of applicability of the peak-decay method. In each case, the criteria given as Equations 45 to 47 will be used to establish the necessary experimental conditions.

1. DNP-Antibody Kinetics

From the earlier calculations

$$k_{-3} = 500 \text{ sec}^{-1}$$

$$k_{-1} = 240 \text{ sec}^{-1}$$

Clearly, this does not meet the criterion of Equation 45 and so k_{-3} cannot be measured.

2. Con A — Sugar Kinetics

From the earlier calculations

$$k_{-3} = 6.2 \text{ sec}^{-1}$$

$$K_3 = 8.7 \times 10^3 \, M^{-1}$$

$$K_2 = 3.3 \times 10^3 \, M^{-1}$$

$$k_{-1} = 240 \text{ sec}^{-1}$$

Since $k_{-1}/k_{-3} = 39$, the criterion of Equation 45 is met. From Equation 46, if $V_e = V_p$, $t_m \leq 0.06$ sec. At a flow rate of 1 mℓ/min, this corresponds to a void volume of only 1.0 $\mu\ell$. Therefore, a higher flow rate (10 mℓ/min) will be used to increase V_m to 10 $\mu\ell$. This would correspond to a column 0.4 × 0.2 cm.

If the immobilized-Con A is equivalent to 1/10 monolayer, the earlier calculations show $m_L/V_m = 150$ nmol/mℓ. The criterion of Equation 47, combined with Equation 15, will establish the competing-inhibitor concentration. The result of the calculation is $[I] \geqslant 3.7 \times 10^{-3}$ M. Note, however, that when $[I] = 0$, $k' = 1.3$. This is so small that most of the PNPM would be lost when the column was washed prior to elution with MDM. Therefore, one must use a higher ligand density or surface area support to give, e.g., $m_L/V_m = 1500$ nmol/mℓ and $[I] \geqslant 0.039$ M. The value of k' for PNPM in the absence of inhibitor is then 13.

The number of moles of ligand in the column will be 15 nmol. Since k' is still rather low, assume that after washing out excess PNPM, 5 nmol PNPM remain in the column. After introducing the inhibitor, 95% of the PNPM should elute within three half-lives, i.e., $3 \ln 2/k_{-3} = 0.34$ sec or 56 $\mu\ell$ at a flow rate of 10 mℓ/min. The average concentration of PNPM in the detector would be 8.9×10^{-5} M.

The major problem with this experiment would probably be column instability due to the high flow rate and small column size. Another problem would be extracolumn band-broadening. One would probably need to use a 1 $\mu\ell$ flow cell.

Comparing these results with the earlier isocratic calculations, it can be seen that the peak-decay experiment should be detecting, on the average, a 100-fold higher concentration of PNPM. Other advantages of the peak-decay method include lack of any need to exactly measure k', u, V_p, V_e, and k_{-1} and the lack of problems related to calculation of H_m and H_{sm} as a function of k'. The advantage of the isocratic method is that it utilizes more ordinary chromatographic conditions and should have fewer technical problems.

3. Protein A — IgG Kinetics

As before, assume that fluorescent-labeled IgG is the solute and unlabeled IgG is the inhibitor

$$k_{-3} \sim 10^{-3} \text{ sec}^{-1}$$

$$K_2 = K_3 = 4 \times 10^8 \ M^{-1}$$

$$k_{-1} = 2.4 \text{ sec}^{-1}$$

Since $k_{-1}/k_{-3} = 2400$, the criterion of Equation 45 is met. From Equation 46, $t_m \leqslant 400$ sec. This is easily obtained. Choosing $F = 0.1$ mℓ/min to keep eluent usage low, $V_m \leqslant 0.67$ mℓ. A 5×0.46 cm column ($V_m = 0.66$ mℓ) would be suitable.

The criterion given by Equation 47 requires $k' \leqslant 0.1$. As was done in the isocratic calculations, assume that the maximum practical IgG inhibitor concentration is 1 mg/mℓ = 6.7×10^{-6} M. Then from Equation 15, $m_L/V_m \leqslant 0.67$ nmol/m$_L$. In the absence of any inhibitor, $k' = 2700$.

If all 0.44 nmol of ligand are filled with IgG molecules and if most of the IgG elutes in three half-lives (35 min or 3.5 mℓ), the average concentration of IgG in the detector will be 1.3×10^{-7} M. This is 10-fold higher than in the isocratic method but would still require a sensitive detection system.

The major advantage of the peak-decay method compared to the isocratic method is that it would take less than 1 hr per run compared to over 14 hr per run for the isocratic method. For complexes with very small k_{-3}, it is apparent that the chromatographic conditions are more "normal" when the peak-decay method is used than when the isocratic method is used.

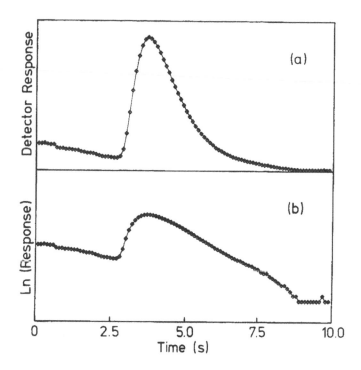

FIGURE 19. Experimental peak-decay curve (a) and natural logarithmic plot (b) for MUM eluted from an immobilized-Con A column at 10 mℓ/min.

D. Experimental Data

1. Immobilized-Con A Data

The experimental system used consisted of Con A immobilized on a 5 μm-diameter diol-bonded silica matrix.[34] The sugar MUM was used as the solute and mannose was used as the inhibitor. Since the dissociation-rate constant of MUM was smaller than that of PNPM and the binding constant was larger, it was possible to use a column of larger volume (0.64 × 0.21 cm, $V_m \sim 18$ μℓ) than was indicated by the example calculations.

The column was initially saturated with MUM at a low flow-rate. The excess MUM was then washed out with buffer (but not too extensively since the MUM bled off of the column fairly rapidly). Finally, 0.1 M mannose was applied to the column at a high flow-rate, and the elution of MUM was monitored using an absorbance detector. Figure 19 shows data from a typical run. The logarithmic plot (Figure 19b) is fairly linear, at least until the tail of the peak where noise and error in locating the baseline begin to cause problems.

To ensure that the criterion given in Equation 46 was obeyed, flow rates from 2 to 10 mℓ/min were tested (Figure 20). Above 8 mℓ/min, the apparent-rate constant reached a plateau value. This trend was predicted by Figure 17. However, the experimental data appeared to be more sensitive to flow rate than the simulated data was; this may be related to band-spreading of the inhibitor front or MUM in the interparticle volume of the column.

To ensure that the criterion given in Equation 47 was obeyed, various concentrations of the inhibitor were used to elute the MUM. Figure 21 shows that as the inhibitor concentration increased (and k′ decreased) the apparent-rate constant reached a plateau value, as expected from Figures 16 and 17.

Since the criteria given as Equations 45 to 47 were based on computer simulations which only modeled some of the processes taking place in real columns, plots like Figures 20 and 21 should be made to ensure that reliable results are obtained.

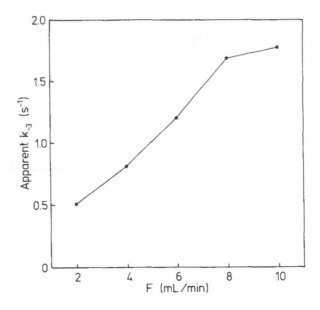

FIGURE 20.   Experimental rate constants from the peak-decay method as a function of flowrate for MUM eluted from an immobilized-Con A column.

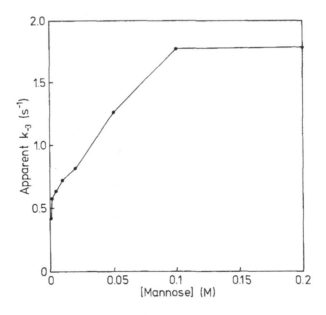

FIGURE 21.   Experimental rate constants from the peak-decay method as a function of competing inhibitor concentration for MUM eluted from an immobilized-Con A column.

Based on the data above, k_{-3} for the Con A - MUM complex was determined to be 1.8 sec^{-1}. This is reasonably close to the literature value[21] of 3.4 sec^{-1} and the isocratically determined[18] value of roughly 2 sec^{-1}.

In contrast to the isocratic experiments, the results from the peak-decay method were very precise (about $\pm$ 5% run-to-run reproducibility and $\pm$ 10% column-to-column reproducibility). This fact alone makes the peak-decay method superior to the isocratic method.

The main difficulties with the peak-decay method appear to be the need for fairly high competing-inhibitor concentrations and the problems associated with detection of the solute at the same time as a concentrated plug of inhibitor passes through the flow cell. Another important factor is the time constant of the detector.

2. Future Prospects for Peak-Decay Measurements

The development of more efficient matrices will increase the usefulness of this method in the future. A particularly interesting application of the method may be in the kinetic evaluation of nonspecific elution methods. For example, if both acidic solutions and organic solvents denature a particular affinity ligand-solute complex, it would be of fundamental and practical interest to determine the apparent k_{-3} value for each eluent and to compare each with the value obtained from competitive elution. This would provide another means of optimizing the chromatographic conditions.

VI. GLOSSARY

Symbol	**Definition**
k'	Capacity factor; the number of column volumes the solute is retained (Equation 14)
L_{col}	Column length
t_r, V_r	Retention time or volume of a solute measured from the time of injection; the first statistical moment of the peak
t_m, V_m	Void time or volume; the retention time of a nonretained solute
V_e	Excluded volume; the elution volume of an excluded, nonretained solute
V_p	Pore volume; $V_p = V_m - V_e$
V_{col}	Empty column volume
F	Volumetric flowrate of the mobile phase
u	Linear velocity; the velocity of the solvent; $u = L_{col}/t_m$
u_e	Linear velocity of an excluded solute; $u_e = uV_m/V_e$
σ^2	Peak variance; the second statistical moment of a peak
N	Number of theoretical plates in the column (Equation 1)
H	Plate height (Equation 2)
E	Solute applied to the column
L	Immobilized affinity ligand
I	Inhibitor or competing ligand present in the mobile phase to control retention of E
K_3, k_3, k_{-3}	Equilibrium and rate constants for binding of solute to ligand
K_2, k_2, k_{-2}	Equilibrium and rate constants for binding of inhibitor to solute (normal-role) or affinity ligand (reversed-role)
k_1, k_{-1}	Rate constants for diffusion of solute into and out of the stagnant mobile phase; $k_1/k_{-1} = V_p/V_e$
m_L	Number of moles of active ligand in the column
m_L/V_p	Effective molar concentration of active ligand
A	Surface area of the matrix in the column
k_3*	First order adsorption rate constant (Equation 12)
d_p	Particle diameter
D_m	Diffusion coefficient of the solute in bulk mobile phase
γ	Tortuosity factor to correct for slower diffusion inside a porous matrix
f	Free or nonretained fraction of solute in the split-peak method

REFERENCES

1. **Giddings, J. C.**, *Dynamics of Chromatography*, Marcel Dekker, New York, 1965.
2. **van Deemter, J. J., Zuiderweg, F. J., and Klinkenberg, A.**, *Chem. Eng. Sci.*, 5, 271, 1956.
3. **Kennedy, G. J. and Knox, J. H.**, *J. Chromatogr. Sci.*, 10, 549, 1972.
4. **Horvath, C. and Lin, H.-J.**, *J. Chromatogr.*, 126, 401, 1976.
5. **Horvath, C. and Lin, H.-J.**, *J. Chromatogr.*, 149, 43, 1978.
6. **Katz, E. D. and Scott, R. P. W.**, *J. Chromatogr.*, 270, 51, 1983.
7. **Katz, E., Ogan, K. L., and Scott, R. P. W.**, *J. Chromatogr.*, 270, 29, 1983.
8. **Stout, R. W., DeStefano, J. J., and Snyder, L. R.**, *J. Chromatogr.*, 282, 263, 1983.
9. **Karger, B. L., Snyder, L. R., and Horvath, C.**, *An Introduction to Separation Science*, John Wiley & Sons, New York, 1973.
10. **Snyder, L. R. and Kirkland, J. J.**, *Introduction to Modern Liquid Chromatography*, John Wiley & Sons, New York, 1979.
11. **Hethcote, H. W. and DeLisi, C.**, *J. Chromatogr.*, 240, 269, 1982.
12. **DeLisi, C., Hethcote, H. W., and Brettler, J. W.**, *J. Chromatogr.*, 240, 283, 1982.
13. **Hethcote, H. W. and DeLisi, C.**, *J. Chromatogr.*, 248, 183, 1982.
14. **Hethcote, H. W. and DeLisi, C.**, in *Affinity Chromatography and Biological Recognition*, Chaiken, I. M., Wilchek, M., and Parikh, I., Eds., Academic Press, Orlando, 1983, 119.
15. **Walters, R. R.**, *J. Chromatogr.*, 249, 19, 1982.
16. **Rakowski, A.**, *Z. Phys. Chem.*, 57, 321, 1906.
17. **Anderson, D. J. and Walters, R. R.**, *J. Chromatogr. Sci.*, 22, 353, 1984.
18. **Anderson, D. J. and Walters, R. R.**, *J. Chromatogr.*, 376, 69, 1986.
19. **Young, M. E., Carroad, P. A., and Bell, R. L.**, *Biotechnol. Bioeng.*, 22, 947, 1980.
20. **Pecht, I. and Lancet, D.**, *Mol. Biol., Biochem. Biophys.*, 24, 306, 1977.
21. **Clegg, R. M., Loonteins, F. G., Van Landschoot, A., and Jovin, T. M.**, *Biochemistry*, 20, 4687, 1981.
22. **Hage, D. S., Walters, R. R., and Hethcote, H. W.**, *Anal. Chem.*, 58, 274, 1986.
23. **Lindmark, R., Biriell, C., and Sjoquist, J.**, *Scand. J. Immunol.*, 14, 409, 1981.
24. **Kasche, V., Buchholz, K., and Galunsky, B.**, *J. Chromatogr.*, 216, 169, 1981.
25. **Chaiken, I. M.**, *Anal. Biochem.*, 97, 1, 1979.
26. **Nilsson, K. and Larsson, P.-O.**, *Anal. Biochem.*, 134, 60, 1983.
27. **Muller, A. J. and Carr, P. W.**, *J. Chromatogr.*, 184, 33, 1984.
28. **Swaisgood, H. E. and Chaiken, I. M.**, this book.
29. **Moore, R. M. and Walters, R. R.**, unpublished results, Iowa State University, 1985.
30. **Giddings, J. C. and Eyring, H.**, *J. Phys. Chem.*, 59, 416, 1955.
31. **Roy, S. K., Weber, D. V., and McGregor, W. C.**, *J. Chromatogr.*, 303, 225, 1984.
32. **Sportsman, J. R. and Wilson, G. S.**, *Anal. Chem.*, 52, 2013, 1980.
33. **Walters, R. R.**, *Anal. Chem.*, 55, 1395, 1983.
34. **Moore, R. M. and Walters, R. R.**, *J. Chromatogr.*, 384, 91, 1987.
35. **Wade, J. L., Bergold, A. F., and Carr, P. W.**, *Anal. Chem.*, 1987, in press.

Chapter 4

QUANTITATIVE CONSIDERATIONS OF CHROMATOGRAPHY USING IMMOBILIZED BIOMIMETIC DYES

Earle Stellwagen and Yin-Chang Liu

TABLE OF CONTENTS

I. INTRODUCTION

During the last decade immobilized dyes have been used to great advantage in the purification of proteins from biological sources. This popularity results from at least three considerations. First, immobilized dyes function as general ligands, i.e., they retain a variety of proteins which can be individually eluted by judicious selection of a competitive biospecific ligand. Recent reviews[1-5] of immobilized-dye chromatography provide references to some of the approximately 500 different proteins purified using this technology. Such proteins do not represent a nested set of closely related proteins but rather range widely across the biological spectrum from bacterial-restriction endonucleases[6,7] to plant flavokinase.[8] Second, immobilized-dye columns can be generated easily and inexpensively. Since dyes are prepared in bulk for the textile industry, the cost for generation of even a commercial-scale[3,9] column represents a modest investment. In addition, many of the dyes advantageous for protein purification are reactive dyes which require neither an activated matrix nor a spacer group for functional-covalent immobilization. Third, and most important, dyes are nonbiodegradable. In contrast to immobilized-biospecific ligands, immobilized dyes are not degraded by the hydrolytic enzymes abundantly present in all crude cellular extracts. Accordingly, an immobilized dye column can be used as the first step in protein purification, facilitating the full expression of the purification enhancement achievable by general ligand-affinity chromatography. Indeed, examples have been reported of purification of proteins in a crude extract to homogeneity in a single chromatographic step using immobilized-dye columns.[10-12] Procedures have also been described to perform immobilized-dye chromatography in the presence of the nonionic detergents necessary to extract and maintain the biofunctionality of particulate proteins.[13]

The reactive triazine dyes have been found to be particularly convenient for the preparation of immobilized-dye columns for protein purification. Such dyes include the monochlorotriazinyl dyes designated Procion H by ICI and Cibacron by Ciba-Geigy and the dichlorotriazinyl dyes designated Procion MX by ICI. These dyes cover the visible spectrum by virtue of coupling of the triazine with anthraquinone, azo, or phthalocyanine chromophores. While all immobilized triazine dyes have been found to be useful in protein-purification procedures, several reports document the selective advantage of one or more immobilized triazine dyes for the purification of a particular protein.[14-16] The blue anthraquinone monochlorotriazine dye, designated by the generic name reactive blue 2, color index 61211, and by the commercial names Cibacron blue F3GA and Procion blue HB, has been most commonly used both for protein purification and for detailed studies of dye:protein interaction. The structure of reactive blue 2 is shown in Figure 1. The popularity of this dye results in large measure from the presumed arbitrary selection of Cibacron blue F3GA in the synthesis of a colored high-molecular-weight polymer for determination of the excluded volume of molecular-sieving columns, a polymer named blue dextran by Pharmacia. The anomolous behavior of some proteins cochromatographed with blue dextran in solvents of low ionic strength led to the independent deduction by Kopperschläger et al.[17] and Haeckel et al.[18] that such proteins have a high affinity for the blue chromophore. Had the Pharmacia chemists selected another triazine dye, the attention of biochemists would likely be directed toward columns of a different hue.

This discussion considers the quantitation of triazine dye:protein interactions using both mobile- and immobilized- reactive blue 2. Emphasis is placed on recent experimentation, both published and unpublished, describing modulation of the affinity, accessibility, and valence of immobilized reactive blue which can be used to advantage in protein purification. It is anticipated that this discussion contains observations which pertain not only to other immobilized triazine dyes but to immobilized ligands in general.

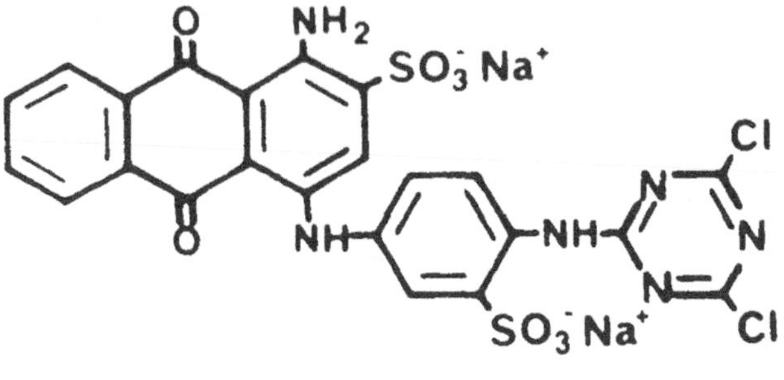

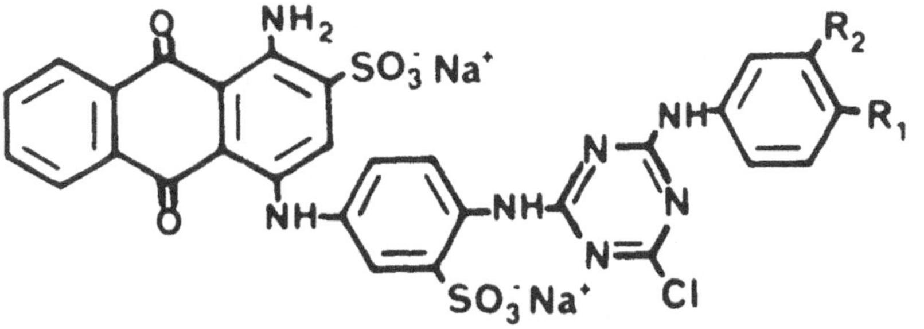

FIGURE 1. Chemical structure for the dichlorotriazine precursor of reactive blue 2, upper structure, and for reactive blue 2, lower structure, at neutral pH. One of the R groups of reactive blue 2 is a sulfonate and the other is a hydrogen depending upon whether the dye is the meta- or para-isomer.

II. INTEGRITY OF COMMERCIAL DYE PREPARATIONS

Most investigators who have examined the properties of reactive blue 2 or prepared and characterized immobilized-reactive blue 2 columns have utilized commercially available samples of the dye. Alternatively, some investigators either have used commercially available immobilized reactive blue 2 columns or have prepared immobilized blue dextran columns using commercially available blue dextran. A noteable exception to this was the limited availability of presumed homogeneous samples of Cibacron blue F3GA from the research laboratories of Ciba-Geigy.

Investigation of the homogeneity of the commercially available samples of Cibacron blue F3GA by various thin-layer chromatographic protocols[19-21] generally indicate a major blue component and minor amounts of dyes of various hues. However, Hanggi and Carr[22] have recently demonstrated by paired-ion reverse-phase HPLC measurements that the Cibacron blue F3GA offered by three major commercial suppliers is either largely or totally devoid of a dye having the structure of reactive blue 2 shown in Figure 1. Typical HPLC elution profiles are shown in Figure 2. The major component is the dichlorotriazinyl precursor of reactive blue 2 whose structure is also shown in Figure 1. This precursor is known commercially as Procion blue MX3G. However, paired-ion reverse phase HPLC has also shown that Procion blue MX3G does not contain material having the structure of the dichlorotriazinyl precursor shown in Figure 1. Since the reactive blue 2 prepared by ICI as Procion blue HB is probably prepared from Procion blue MX3G, it is likely that neither Cibacron blue F3GA or Procion blue HB contains reactive blue 2 as a major component. Fortunately, Hanggi and Carr[22] have described detailed procedures for the synthesis and purification of reactive blue 2 so that dye function can be more reliably assigned to dye structure in future measurements. Such procedures are particularly timely since Cibacron blue F3GA is no longer being produced.

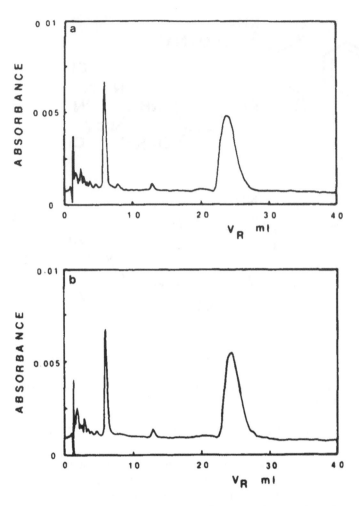

FIGURE 2.   Elution profiles for samples of commercial preparations of reactive blue 2 obtained using reversed-phase paired-ion high-pressure liquid chromatography. *The uppermost profile was observed using Sigma* reactive blue 2, the middle profile using Polyscience Cibacron blue F3GA, and the lower profile using Pierce Cibacron blue F3GA. Samples of dye were injected into a 10-cm Hypersil, 5-μm ODS column equilibrated with 2.25 m*M* tetrabutylammonium chloride and 4.5 m*M* phosphate buffer, pH 7.0, in 55% methanol. The equilibration solvent was flowed at 1 mℓ/min and the effluent absorbance measured at 254 nm. The authentic dichlorotriazine precursor of reactive blue 2 has a retention time of 25 mℓ while the meta- and para-isomers of authentic-reactive blue 2 have elution times of 10.4 and 9.6 mℓ, respectively. (From Hanggi, D. and Carr, P., *Anal. Biochem.*, 149, 91, 1985. With permission.)

III. NONCOVALENT POLYMERIZATION

Since reactive blue 2 is a polyaromatic compound it would be expected to exhibit some self-association in solution by ring stacking. The dependence of the visible-absorbance spectrum on dye concentration in neat water confirms such expectation as shown in Figure 3. These concentration-dependent spectral perturbations are consistent with a two-state transition between the monomeric dye and an alternative form presumed to be a ring-stacked polymer. It should be noted the appearance of the alternative form can be detected at quite low-dye concentration, beginning at about 5 μ*M*. Since concentrated solutions of dye remain

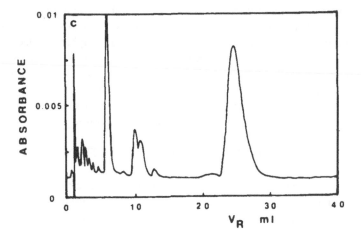

FIGURE 2C (continued)

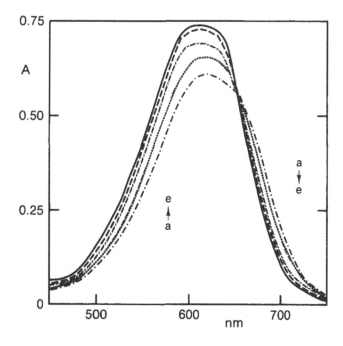

FIGURE 3. The effect of dye concentration on the visible spectrum in water. The dye was a purified sample of Cibacron blue F3GA obtained from the research laboratories of Ciba-Geigy. The dye concentration in solutions a through e is 0.5, 5, 50, 500, and 5000 μM, respectively. All spectra were obtained on the same chart tracing by employing cells with different optical-path lengths.

soluble in neat water, the presumed stacked polymers are soluble. However, if salt is added to solutions of concentrated dye a tacky, colored precipitate quickly appears, suggesting that the salt masks the charge repulsions between dispersed stacked polymers in neat water.

Unfortunately, the protocols[23-25] commonly used for the conjugation of reactive blue 2 to chromatographic matrices are based on a protocol used in the textile industry for staining cellulose cloth. Concentrated salts are added to dye solutions to initiate adsorption of the dye to cloth to promote preferential reaction of the triazine chloride with cellulose as opposed to the solvent water. Such a protocol leads to a high-local concentration of dye on the cloth and by analogy on chromatographic matrices. Accordingly, it is probably erroneous to assume that dye is randomly distributed on the surface of chromatographic matrices when covalent coupling is performed at high-dye concentration and high-ionic strength. Any locally high concentrations of immobilized dye would have a marked effect on the multivalency of dye columns when chromatographing oligomeric proteins as described below. In order to achieve a more random distribution of immobilized dye for future quantitative measurements, it might be well to minimize both dye and salt concentration during immobilization.

IV. DYE:PROTEIN INTERACTION

Since reactive dyes are designed to physically adsorb to textile surfaces, it might be anticipated that dyes would also bind extensively and indiscriminately to protein surfaces. While such indiscriminate binding to protein surfaces may occur with very low affinity, many proteins possess a very limited number of high affinity sites for reactive blue 2. The total number of protein sites for noncovalent binding of dye can be most easily determined by spectrophotometric titration.[26] Adsorption of dye to a protein surface decreases the polarity of the dye environment relative to water. This change in polarity causes a red shift in the absorbance spectrum of the dye generating a difference spectrum similar to that produced by ethylene glycol as shown in Figure 4. For many proteins, the difference-absorbance intensity displays a hyperbolic dependence on dye concentration indicative of saturation binding as shown in Figure 5A. Analysis of such dependence is frequently consistent with the complexation of 1 dye/protein-globular domain having an affinity constant ranging between 10^5 and 10^7 M^{-1}. Precise evaluation of affinity constants for mobile dye:protein complexes provide very important reference values for the interpretation of immobilized dye:protein interactions.

The location of the dye-binding site on the protein surface is commonly investigated by competitive measurements with biofunctional ligands. In the case of spectrophotometric measurements, ligands competitive with the dye site will cause the loss of the difference spectrum associated with dye binding. A typical example is shown in Figure 5B, where the loss of the difference spectrum is appropriate to the dissociation constants for the competing ligands NAD or NADH. Alternatively, competitive-bifunctional measurements can identify the dye-binding site on the protein surface. If the dye-binding site is an enzyme-catalytic site, then the dye should act as a competitive inhibitor upon double-reciprocal analysis of catalytic rates. A typical example is illustrated in Figure 6. In such measurements, the competition-inhibition constant is equated with the dissociation constant for the mobile dye:enzyme complex. If the dye-binding site is an allosteric site, then the dye should either itself function as an allosteric effector or else prevent the expression of the biospecific effector. If the binding site is the biofunctional site of a receptor or binding protein, the dye should compete with the biospecific ligand for binding.

Quantitative measurements of mobile dye:protein interactions not only have demonstrated the selectivity of dye:protein interaction but also have revealed that the dye often binds with greater affinity than does the biospecific ligand. Clearly, the dye and the biospecific ligand often share many features including size, atoms of similar polarity, presence of one or more aromatic rings, and the presence of formal-negative charges. Apparently, the dye has sufficient flexibility to both occupy biospecific sites and to participate in an array of noncovalent forces whose cumulative strength is at least equivalent to the strength of the forces involved in the complexation of the biospecific ligand. Since the dye can mimic the biospecific ligand, the term biomimetic dye seems particularly apt.

The complex between Cibacron blue F3GA and crystalline horse-liver alcohol dehydrogenase has been examined[27] at modest resolution using the 0.37 difference electron-density

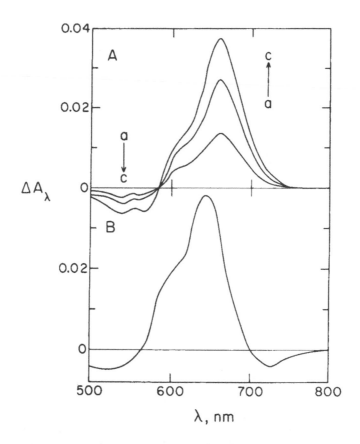

FIGURE 4. Changes in the visible absorbance spectrum of reactive blue 2 caused by addition of either lactate dehydrogenase or ethylene glycol. The dye was a purified sample of Cibacron blue F3GA obtained from the research laboratories of Ciba-Geigy. Panel A illustrates the difference spectra observed following addition of increasing concentrations of dye to a 3 μM solution of tetrameric-rabbit muscle lactate dehydrogenase in 10 mM Tris-HCl buffer, pH 7.5, containing 0.5 mM mercaptoethanol. The sample cuvet contained the protein solution and the reference cuvet the protein solvent. Difference spectra, a, b, and c, were obtained following addition of 2.15, 4.28, and 7.47 μM dye, respectively. Panel B illustrates the difference spectrum observed following addition of 50% ethylene glycol to a solution of dye. Both the sample and reference cuvets contained 21.5 μM dye. (From Thompson, S. T. and Stellwagen, E., *Proc. Natl. Acad. Sci. U.S.A.*, 73, 361, 1976. With permission.)

map shown in Figure 7. This difference map indicates that the para isomer of the dye selectively binds to the coenzyme-binding site of the enzyme in an elongated orientation similar to but not identical with the coenzyme. The anthraquinine (D) ring of the dye occupies the pocket which binds the adenine portion of the coenzyme. The internal-phenylsulfonate (C) ring digresses from the structure of the complexed coenzyme presumably to form an electrostatic interaction between the sulfonate and the guanidinium group of Arg 271. The triazine (B) ring is oriented, as is the pyrophosphate moiety of the coenzyme, placing the triazine chloride in contact with the solvent. Such location would predict that a dye immobilized at its triazine ring would bind alcohol dehydrogenase as does mobile dye. The terminal-phenylsulfonate (A) ring is oriented differently than the nicotinamide portion of the coenzyme, presumably to facilitate electrostatic interaction between the terminal sulfonate of the dye with the guanidinium group of Arg 369.

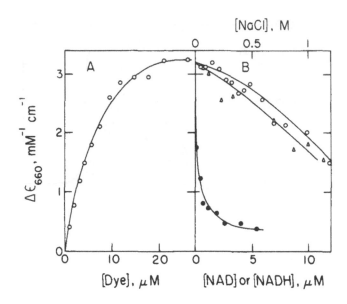

FIGURE 5.   Difference spectral titrations. Panel A illustrates the increase
in the difference spectrum of the dye observed at 660 nm following addition
of increasing concentrations of dye to a 3.0 μM solution of tetrameric-
rabbit muscle lactate dehydrogenase. Panel B illustrates the decrease in
the difference spectrum of the dye observed at 660 nm following addition
of either: ○, NAD: ●, NADH; or △, NaCl to a solution containing 2.9
μM rabbit muscle lactate dehydrogenase and 26.1 μM dye. All solutions
contained 10 mM Tris-HCl buffer, pH 7.5, and 0.5 mM mercaptoethanol.
The dye was obtained from the research laboratories of Ciba-Geigy. (From
Thompson, S. T. and Stellwagen, E., *Proc. Natl. Acad. Sci. U.S.A.*, 73,
361, 1976. With permission.)

Since the orientation of reactive blue 2 in the coenzyme site of alcohol dehydrogenase
appears to be modulated by electrostatic considerations, Lowe and co-workers[28] have syn-
thesized a variety of analogs of reactive blue 2 in which the charge, size and location of
the substituent on the A ring is altered. As shown in Table 1, the affinity between alcohol
dehydrogenase and analogs of reactive blue 2 can be changed by three orders of magnitude
by such alterations. These results provide strong encouragement that the chemical structure
of reactive blue 2 can be altered in productive ways to increase its bioselectivity for a variety
of targeted proteins.

V. ACCESSIBILITY OF IMMOBILIZED DYE

One of the advantages of using immobilized-dye columns is that the intensity of the color
is a good qualitative indication of the amount of dye retained by the matrix. Chambers[29]
has described a procedure to quantitatively measure the total concentration of Cibacron
F3GA conjugated to a matrix by absorbance measurements following acid hydrolysis. The
amount of total-immobilized dye accessible to a given protein can be determined by equilibrium partition measurements using either recycling[30,31] or centrifical procedures[32] or by
frontal-chromatographic analysis.[32] A typical example of a frontal analysis is shown in Figure
8 which is plotted in the form

$$\frac{1}{V_L - V_o} = \frac{E_o}{DV_i} + \frac{1}{DV_iK_D^A} \qquad (1)$$

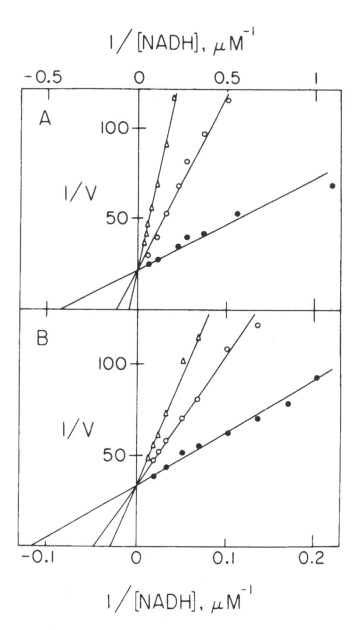

FIGURE 6. The inhibition of lactate dehydrogenase catalysis by reactive blue 2. Panel A illustrates the catalytic rate of the rabbit-muscle enzyme in the presence of ●, no dye; ○, 0.27 μM dye, and △, 0.86 μM dye. Panel B illustrates the catalytic rate of the bovine-heart enzyme in the presence of ●, no dye; ○, 2.15 μM dye; and △, 6.83 μM dye. The inhibition constant for the dye is 0.13 μM with the muscle enzyme and 2.1 μM with the heart enzyme. All measurements were obtained at 25°C using 10 mM Tris-HCl, pH 7.5, as the assay buffer. The initial rate of catalysis, V, has the units change in 340 nm absorbance/min. The dye was obtained from the research laboratories of Ciba-Geiry. (From Thompson, S. T. and Stellwagen, E., *Proc. Natl. Acad. Sci. U.S.A.*, 73, 361, 1976. With permission.)

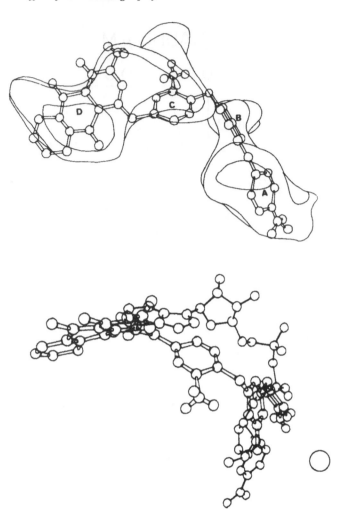

FIGURE 7.   The binding of the para-isomer of Cibacron blue F3GA to the coenzyme-binding site of horse-liver alcohol dehydrogenase as determined by X-ray diffraction measurements. The enzyme was crystallized in 50 mM Tris-HCl buffer, pH 8.4, containing 2-methyl-2,4-pentane diol. The crystalline protein was then dialyzed at 4°C against 17 μM solution of the dye in the crystallization solvent. The purified isomer of the dye was obtained from Ciba-Geigy. The upper portion of the figure illustrates the fitting of the dye structure to a contoured 0.37 nm-difference electron-density map. The lower portion of the figure compares the orientation of the dye (open bonds) with that of NADH (closed bonds) in the coenzyme-binding site of the enzyme as deduced from crystallographic measurements. (From Biellmann, J-F., Samama, J. P., Brandén, C. I., and Eklund, H., *Eur. J. Biochem.*, 102, 107, 1979. With permission.)

Such measurements are most reliable if the interaction between protein and immobilized dye is monovalent. This situation pertains if a protein having a single-dye complexation site is employed or if a matrix having a relatively low total concentration of immobilized dye is used. In instances where multivalent interactions are either known or anticipated to occur, the concentration of accessible-immobilized dye can be estimated by graphical analysis of equilibrium-partition measurements.[31]

We have found[32] that for a lighly conjugated Sepharose CL-4B matrix containing a total

Table 1
REACTIVE BLUE 2 ANALOGS

Substituent on A ring	K_D^D (μM)		
	ortho	meta	para
–COO$^-$	0.2	0.06	5.9
–SO$_3^-$	0.4	1.6	9.3
–OPO$_3$H$^-$			10.5
–N(CH$_3$)$_3$$^+$			172.0

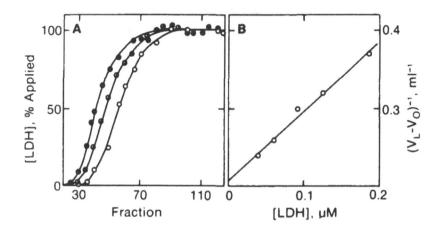

FIGURE 8. Measurement of the concentration of accessible-immobilized dye by frontal chromatography. Panel A illustrates typical elution profiles observed following application of: ○, 0.040 μM lactate dehydrogenaes; ○, 0.125 μM lactate dehydrogenase; and ●, 0.186 μM lactate dehydrogenase; Panel B illustrates the dependence of the midpoint of the frontal-elution profiles on the concentration of enzyme applied plotted according to equation 1. All measurements were obtained in 100 mM phosphate buffer, pH 7.5 and 23°C, using a Sepharose CL-6B containing a total concentration of 0.12 mM immobilized Polyscience Cibacron blue F3GA. (From Liu, Y. C., Ledger, R., and Stellwagen, E., *J. Biol. Chem.*, 259, 3796, 1984. With permission.)

concentration of 0.12 mM immobilized-Cibacron blue F3GA, only 2.1 ± 0.3 μM of immobilized dye is accessible to tetrameric rabbit-muscle lactate dehydrogenase. Quantitative analysis using both zonal- and frontal-chromatographic procedures indicated that the lactate dehydrogenase:immobilized dye interaction on this column is monovalent. The low-fractional accessibility of immobilized dye is not limited to lactate dehydrogenase. As shown in Figure 9, no more than 5% of the immobilized dye on the 0.12 mM column is accessible to a variety of proteins. While the percentage-total dye accessible to these proteins exhibits an inverse dependence on their molecular weight, such dependence is quite distinct from the partition coefficient of these proteins on Sepharose CL-6B as shown in Figure 9. Thus the low-fractional accessibility of immobilized dye is not primarily due to the exclusion of protein from a gel volume accessible to the small dye prior to conjugation.

In searching for a rationale to explain the low accessibility of immobilized Cibacron blue F3GA, we observed that the fractional accessibility of dye to lactate dehydrogenase increased as the concentration of phosphate buffer is diminished. The accessibility was increased to a value commensurate with its partition value as shown in Figures 9 and 10. Since the mobile dye is designed to absorb to surfaces such as the Sepharose matrix, we reasoned that high concentrations of phosphate buffer masked the three anionic-sulfonate charges, facilitating

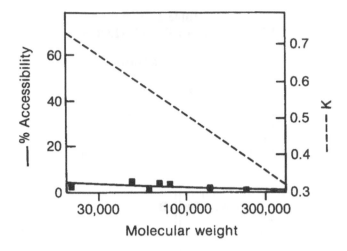

FIGURE 9.   Comparison of the percentage accessibility of proteins to Cibacron blue F3GA immobilized on Sepharose CL-6B with the partition coefficient, K, of proteins on Sepharose CL-6B. Polyscience Cibacron blue F3GA was immobilized through its triazine ring using the procedure of Heyns and de Moor.[24] The accessibility of protein was measured by frontal analysis using a column containing a total concentration of 0.14 mM immobilized dye. The column was equilibrated with 100 mM acetate buffer, pH 5.5, for measurements with phosphoglycerate kinase and pyruvate kinase; 100 mM Tris-HCl buffer, pH 7.5, for measurements with malate dehydrogenase; and 100 mM phosphate buffer, pH 7.5, for measurements with the remaining proteins. The percentage accessibility was calculated assuming that each protein was retained by a single-immobilized dye. The experimental values for percentage accessibility shown as filled squares and reading from left to right represent measurements obtained using rabbit-muscle adenylate kinase, rabbit-muscle phosphoglycerate kinase, porcine-heart isocitrate dehydrogenase, pig-heart mitochondrial malate dehydrogenase, horse-liver alcohol dehydrogenase, rabbit-muscle lactate dehydrogenase, rabbit-muscle pyruvate kinase, and bovine-liver glutamate dehydrogenase.

absorption of the dye to the matrix surface. As the concentration of phosphate buffer is diminished, charge repulsions cause the dye to desorb from the matrix surface and extend into the solvent where the dye is accessible to protein.

We attempted[33] to model this situation by observing the interaction between mobile Cibracon blue F3GA and unconjugated or white Sepharose CL-6B. The mobile dye used in this model study had been converted to the monohydroxytriazine derivative in order to prevent covalent attachment of the dye to white Sepharose. We observed that the dye rapidly eluted from a white-Sepharose column after zonal application of the dye to a column equilibrated with neat water, as shown in Figure 11. However, as the concentration of phosphate buffer was increased in the column-equilibration solvent, the emergence of dye zone from the column was increasingly delayed, reaching a maximal delay in about 200 mM phosphate buffer as shown in Figure 11. These measurements indicate that the retention of mobile dye by white Sepharose is directly dependent upon the concentration of phosphate buffer while the exposure of immobilized dye on blue Sepharose is inversely dependent upon phosphate concentration. The dependence of each of these phenomena on phosphate concentration describe a common relationship as shown in Figure 10.

All salts are not equivalent, even when normalized by ionic strength, in their ability to delay the emergence of the dye zone from white-Sepharose columns. As shown in Figure 12, the maximal-elution volume for the dye zone varied from about 2 to 6 mℓ when the

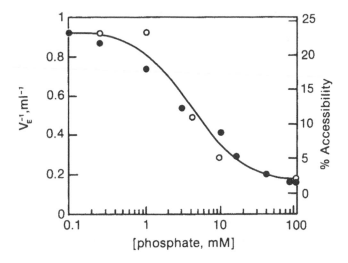

FIGURE 10. Comparison of the phosphate dependence of the fractional accessibility of immobilized dye to lactate dehydrogenase with the phosphate dependence of the retention of mobile dye by the unconjugated matrix. The accessibility of rabbit-muscle lactate dehydrogenase to Sepharose CL-6B containing 0.14 mM total-immobilized Polyscience Cibacron blue F3GA was measured by a saturation procedure and results are shown as open circles. A 1.0 mℓ aliquot of immobilized dye was equilibrated with a desired concentration of phosphate buffer, pH 7.5. The immobilized dye was then equilibrated with 2.0 mℓ of a very-concentrated solution of enzyme in the same phosphate buffer. The immobilized dye exposed to enzyme was then washed with 10 mℓ of the same phosphate buffer and eluted with a concentrated solution of either the competitive ligand, NADH, or with NaCl. Corrections were made for the retention of enzyme by unconjugated Sepharose CL-6B under the same conditions. Such corrections became substative at low concentrations of phosphate buffer. The filled circles indicate values for the elution volume of free dye obtained following zonal application of 10 μℓ of 16.6-mM dye to a 1.0-mℓ bed volume of unconjugated Sepharose CL-6B equilibrated and developed with phosphate buffer, pH 7.5, of the indicated concentrations. Typical elution profiles obtained using this protocol are shown in Figure 11.

white column was equilibrated with a variety of salt solutions. It should be noted that salts such as thiocyanate and nitrate which have a small maximal-elution volume are particularly effective in destabilizing the structure of bulk water. In Figure 12, the viscosity B coefficient is used as a quantitative index of the effect of salts on water structure. Salts which have a low B coefficient, i.e., which are particularly effective in destabilizing water structure, are also effective in weakening the dye:matrix interaction as indicated by the low-elution volume. Such correlation suggests that at least a portion of the dye:matrix interaction is hydrophobic and accordingly dependent on water structure.

Salt also serves to weaken dye:protein interaction. This interaction was modeled by measurement of the competitive-inhibition constant of mobile dye for the coenzyme-binding site of lactate dehydrogenase as a function of phosphate-buffer concentration. As shown in Figure 13, the inhibition constant increases monotonically with increasing phosphate concentration well above an ionic strength of 0.2 M where electrostatic considerations would be completed. The monotonic function shown in Figure 13 suggests the dye:protein interaction also has a hydrophobic component which is weakened as increasing phosphate concentration decreases water structure.

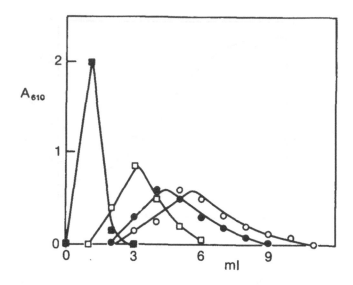

FIGURE 11.   Effect of phosphate concentration on the zonal chromatography of mobile Cibracron blue F3GA. A 10-µℓ solution of Polyscience Cibacron blue F3GA was applied to a 1.0-mℓ bed volume of unconjugated Sepharose CL-6B. The illustrated-elution profiles were obtained using ■, neat water; □, 16 m*M* phosphate buffer; ●, 40 m*M* phosphate buffer and ○, 100 m*M* phosphate buffer for equilibration and development of the column. All measurements were obtained at pH 7.5 and ambient temperature.

These correlations suggest that salts can modulate the interaction between proteins and immobilized dyes by several mechanisms. One mechanism involves masking the negative charges on the immobilized dye to facilitate adsorption of the dye to the matrix surface. This mechanism would modulate both the capacity of an immobilized-ligand column and the potential for multivalent interactions. All salts should be equivalent in this mechanism, when normalized by ionic strength, and exhibit a maximal effect at an ionic strength of 0.2 *M*. A second mechanism concerns the weakening of hydrophobic interactions. This mechanism has at least two opposing effects: a weakening of the immobilized dye:matrix interaction which would favor retention of a protein and a weakening of the immobilized dye:protein interaction which would disfavor retention.[34] All salts are not equivalent in this mechanism and the magnitude of the effect continues with increasing salt concentrations beyond an ionic strength of 0.2 *M*. It is likely that all these phenomena are operational when protein is eluted from immobilized-dye columns.

VI. VALENCE

Thresher and Swaisgood[30] provided the first report of a quantitative investigation of the interaction between immobilized Cibacron blue F3GA and tetrameric lactate dehydrogenase, a model system that was subsequently characterized independently by several other groups. Thresher and Swaisgood immobilized the dye on glycerolpropyl-controlled pore-glass beads and studied its interaction with bovine-heart lactate dehydrogenase by equilibrium partitioning using a microrecirculation reactor. The total concentration of enzyme recirculated through the dye matrix was increased stepwise and the enzyme concentration at equilibrium after each stepwise addition was measured spectrophotometrically using a flow cell. The enzyme concentration measurements were analyzed using Equation 2 which had been previously been developed by Taylor and Swaisgood[35] assuming monovalent interaction.

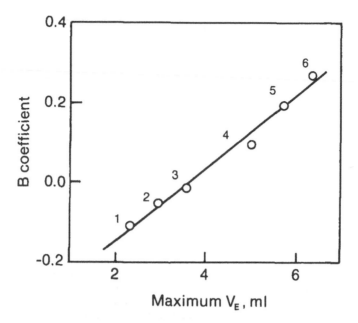

FIGURE 12. Dependence of the maximal-elution volume of mobile Cibacron blue F3GA from unconjugated Sepharose CL-6B on the nature of
salt in the elution solvent. Zonal chromatography using Polyscience Cibacron blue F3GA was done as described in the legend to Figure 11. The
equilibration and development solvent contained 200 mM of each salt
adjusted to pH 7.5 but no phosphate buffer. Prior measurements indicate
that this concentration of salt gives the maximal value for the elution
volume, V_E. The numbered experimental values indicate the following
salts: (1) NaSCN, (2) KNO$_3$, (3) KCl, (4) KF, (5) K$_2$SO$_4$, and (6) potas-
sium-phosphate buffer, pH 7.5. The B coefficient is defined in the equation

$$\eta = 1 + Ac^{1/2} + Bc$$

where η is the viscosity of a solution of salt of concentration "c" relative
to the viscosity of the neat solvent.

$$\left(\frac{E_M}{E_L} - K\right)^{-1} = \frac{K_D^D}{KD} + \frac{E_L}{D} \qquad (2)$$

As shown in Figure 14, the recycling measurement can be fit with a monovalent interaction
in which the partition coefficient for the macromolecule is 1.0, the concentration of accessible
immobilized dye is 29 μM, and the dissociation constant for the immobilized dye:protein
monovalent complex is 2.5 μM. This latter value is in excellent agreement with analysis of
the spectral titration of the protein with mobile dye, which give an average value for the
mobile dye:protein complex of 2.5 ± 0.7 μM. These results indicate that affinity of the
enzyme for Cibacron blue F3GA is not perturbed by conjugation through the triazine ring,
that monovalent interaction between the immobilized dye and the tetrameric protein predominate at an accessible dye concentration of 29 μM, and that the recycling equilibrium
partition procedure provides reliable results.

Liu et al.[31] subsequently reported on the interaction between tetrameric-rabbit muscle
lactate dehydrogenase and Cibacron blue F3GA conjugated with Sepharose CL-6B. This
interaction was examined using three different techniques involving frontal chromatography,

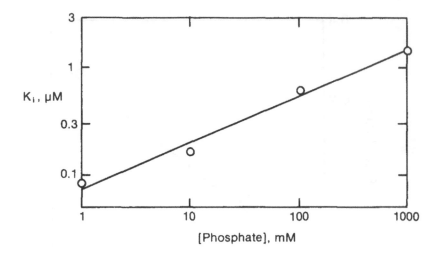

FIGURE 13.   The effect of phosphate-buffer concentration on the inhibition constant of mobile dye for rabbit-muscle lactate dehydrogenase. In each case, the mobile dye, Polyscience Cibacron blue F3GA, competed with NADH for a common form of the enzyme. Measurements were made at ambient temperature and pH 7.5 using the standard-enzymic assay protocol for the enzyme.

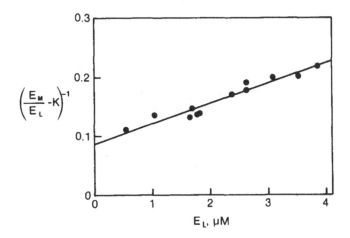

FIGURE 14.   Equilibrium partition analysis of the complexation of Cibacron blue F3GA immobilized onto glycerolpropyl-porous glass with bovine-heart lactate dehydrogenase. Controlled pore-glass beads, 1489 Å pore diameter, were heated with γ-glycidoxy-propyltrimethoxysilane and then reacted with Pierce Cibacron blue F3GA. Results obtained from equilibrium-partition measurements are plotted according to Equation 2. (From Thresher, W. C. and Swaisgood, H. E., *Biochim. Biophys. Acta, 749,* 214, 1983. With permission.)

zonal chromatography and a equilibrium partition. The results obtained using frontal chromatography, shown in Figure 15, were analyzed using Equation 3 introduced by Nichol et al.[36] for monovalent interactions.

$$\frac{1}{V_L - V_o} = \frac{E_o}{DV_i} + \frac{1 + LK_L^A}{DV_iK_D^A} \tag{3}$$

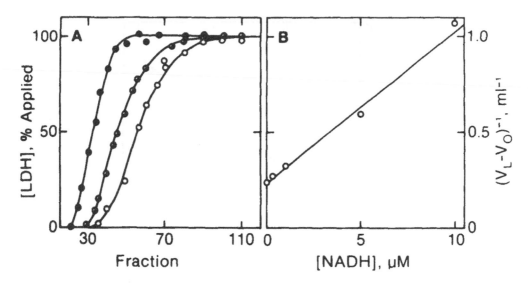

FIGURE 15. Frontal analysis of the complexation of immobilized Cibacron blue F3GA with rabbit-muscle lactate dehydrogenase (LDH). Panel A illustrates elution profiles observed for a 40-nM solution of enzyme in the presence and absence of the competitive mobile ligand NADH: ○, no NADH; ◑, 1 μM NADH; and ●, 5 μM NADH. Panel B illustrates the dependence of the midpoint of such profiles on the concentration of NADH plotted according to Equation 3. The column contained a total of 0.12 mM of immobilized Polyscience Cibacron blue F3GA. Measurements were obtained in 100-mM phosphate buffer, pH 7.5, at 23°C. (From Liu, Y. C., Ledger, R., and Stellwagen, E., *J. Biol. Chem.*, 259, 3796, 1984. With permission.)

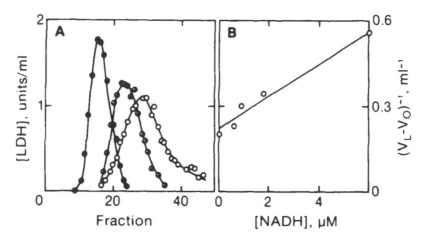

FIGURE 16. Zonal analysis of the complexation of immobilized Cibacron blue F3GA with rabbit-muscle lactate dehydrogenase (LDH). Panel A illustrates typical-elution profiles obtained following zonal application of 100 μℓ of a 1.2 μM solution of enzyme in the presence and absence of the competitive ligand NADH: ○, no NADH; ◑, 0.9 μM NADH; and ●, 6 μM NADH. Panel B illustrates the dependence of the maximum ordinate of each elution profile on the concentration of NADH present, plotted according to Equation 4. The column contained a total of 0.12 mM immobilized Polyscience Cibacron blue F3GA. Measurements were obtained in 100 mM phosphate buffer, pH 7.5, at 23°C. (From Liu, Y. C., Ledger, R., and Stellwagen, E., *J. Biol. Chem.*, 259, 3796, 1984. With permission.)

Typical results obtained using zonal chromatography, shown in Figure 16, were analyzed using Equation 4 which was developed in its general form by

$$\frac{1}{V_L - V_o} = \frac{K_D^D}{D(V_o - V_v)} + \frac{LK_D^D}{DK_L^D(V_o - V_v)} \qquad (4)$$

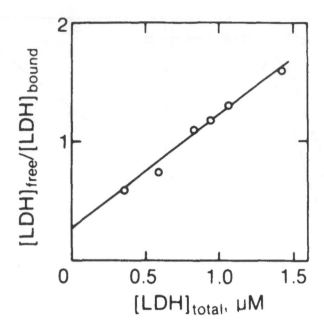

FIGURE 17. Equilibrium partition analysis of the complexation of rabbit muscle lactate dehydrogenase with immobilized Cibacron blue F3GA. A series of 140 μℓ portions of the immobilized Cibacron dye described in the legend to Figure 15 was equilibrated with 250 μℓ of 100 mM phosphate buffer pH 7.5, containing variable initial concentrations of rabbit-muscle lactate dehydrogenase denoted as $[LDH]_{Total}$. The concentration of uncomplexed enzyme in equilibrium with the immobilized dye, denoted as $[LDH]_{free}$, was determined by measurement of the supernatant following centrifugation.

The term $[LDH]_{bound} = [LDH]_{total} - [LDH]_{free}$.

Results are plotted according to Equation 5 in which the ordinate is equivalent to the left-hand term. (From Liu, Y. C., Ledger, R. and Stellwagen, E., *J. Biol. Chem.*, 259, 3796, 1984. With permission.)

Dunn and Chaiken[37] for monovalent interactions. Results obtained by equilibrium partition, shown in Figure 17, were analyzed using Equation 5

$$\frac{E_L(V + V_A)}{E_oV - E_L(V + V_A)} = \frac{E_oV}{DV_A} + \frac{K_D^D(V + V_A)}{DV_A} \tag{5}$$

which was developed in its general form for monovalent interaction by Graves and Wu.[38] Each of these analytical procedures gave a linear relationship indicative of a monovalent interaction at the effective immobilized-dye concentration of 2.0 to 0.3 μM. The dissociation constant for the immobilized dye:lactate dehydrogenase complex obtained from all three procedures compare favorably with each other and with the value for the mobile dye:enzyme complex as shown in Table 2. Such comparisons again indicate that immobilization of the dye through its triazine ring does not perturb its complexation with lactate dehydrogenase and that such interactions are monovalent at a concentration of accessible dye of 2 μM. Table 2 further indicates that the competition between a mobile-biospecific ligand and immobilized dye for enzyme can be correctly evaluated using these procedures. Additionally,

Table 2
COMPARISON OF VALUES

Procedure	D (μM)	K_D^D (μM)	K_L^D (μM)
Immobilized dye			
Frontal analysis	1.7	0.23	2.4
Zonal analysis		0.19	3.7
Equilibrium partition	2.3	0.17	
Mobile dye			
Absorbance		0.49	7.0
Fluorescence			3.5
Catalysis		0.13	3.4

these results indicate that zonal chromatographic procedures can also be used reliably to evaluate immobilized dye interactions.

About the same time, Yon and Kyprianou[39] reported the first evidence for a multivalent interaction between tetrameric rat-liver lactate dehydrogenase and Blue Sepharose-4B obtained from a commercial source. Analysis of their frontal chromatographic measurements at 4°C by Equation 3 yielded a curvilinear relationship suggestive of the presence of multivalent interactions. Yon and Kyprianou introduced an exponential term into Equation 3 to generate Equation 6 which is similar to a Hill equation

$$\frac{1}{V_L - V_o} = \frac{E_o}{DV_i} + \frac{(1 + LK_L^A)^n}{DV_i K_D^A} \tag{6}$$

in which the exponent "n" is the degree of cooperativity. Analysis of their frontal measurements using Equation 6 gave an "n" value of 1.23 indicative of a modest degree of cooperative or multivalent binding of the tetrameric enzyme.

Hogg and Winzor[31] subsequently reported on the interaction of the same Blue Sepharose but now with rabbit-muscle lactate dehydrogenase at 20°C using recycling-equilibrium partition measurements in the presence and absence of the competitive ligand, NADH. Their results, shown in Figure 18, were analyzed using Equation 7 which allows for maximal multivalency but

$$\frac{1 - (E_L/E_o)^{1/f}}{(E_L/E_o)^{1/f}} = \frac{DK_D^A}{1 + LK_L^A} - \frac{fK_D^A E_o[1 - (E_L/E_o)^{1/f}]}{1 + LK_L^A} \tag{7}$$

constrains the immobilized dye:enzyme complex to have a single association constant. The linear display of the observed analytical values indicates that Equation 7 can accommodate the occurrence of multivalent interactions between the immobilized dye and the tetrameric enzyme. The experimental values can be fit with an accessible-immobilized dye concentration of 37 μM, an NADH:enzyme dissociation constant of 12.3 ± 9.6 μM and an immobilized dye:enzyme dissociation constant of 44 ± 6 μM. While the NADH:enzyme dissociation constant is appropriate to independent measurements, the immobilized dye:enzyme dissociation constant is about two orders of magnitude larger than values reported for the mobile dye:enzyme complex[26] and for the immobilized dye:enzyme complex at monovalency.[32] The large dissociation constant for the immobilized dye:enzyme constant appears to result because Hogg and Winzor chose to express the concentration of immobilized dye in terms of the total liquid in the recycling system. Had they chosen to express this concentration in terms of the liquid volume in the gel matrix accessible to the protein, it is anticipated that the dissociation constant for the immobilized dye:enzyme complex would decrease to a value characteristic for the mobile dye:enzyme complex.

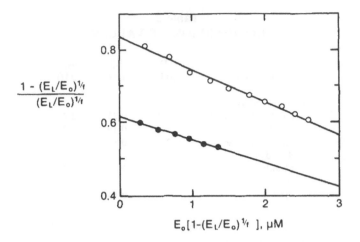

FIGURE 18.   Equilibrium-partition analysis of the complexation of rabbit-muscle dehydrogenase with Blue Sepharose CL-6B. The immobilized dye was purchased from Pharmacia and contained 2.5 μmol dye/mℓ. Increasing concentrations of enzyme were added stepwise to a solvent continuously recirculated past 1.7 g of the immobilized dye. The solvent contained imidazole and NaCl having an ionic strength of 0.4 *M* and a pH of 7.5 and was maintained at 20°C. The concentration of enzyme in the recycling solvent was measured after equilibration of each incremental addition. Results obtained in the presence and absence of competitive mobile ligand, NADH, are plotted in the form of Equation 7: ○, no NADH; ●, 4.3 μ*M* NADH. (From Hogg, P. J. and Winzor, D. J., *Arch. Biochem. Biophys.*, 234, 55, 1984. With permission.)

More recently, Hogg and Winzor[40] have described measurements of the interaction of dimeric-alcohol dehydrogenase which suggest that the microdissociation constants involved in bivalent interactions are not equivalent. Two different amounts of the same blue Sepharose were used in separate recycling-equilibrium partition measurements. Experimental values obtained using the lesser amount of blue Sepharose, 0.5 g, were used to estimate the association constant for the dye:enzyme complex and the concentration of accessible dye assuming monovalency. These estimates were then refined to simultaneously fit the experimental values obtained using both 0.5 and 2.0 g blue Sepharose, allowing for both bivalent interactions and for the nonidentity of the two microassociation constants in the bivalent interaction. The results are shown in Figure 19 in the form of a Scatchard plot using Equation 8 to model the system.

$$r \approx \frac{6\,K_{D1}^A E_L - 1 + [(1 + 2\,K_{D1}^A E_L)^2 + 8\,K_{D1}^A K_{D2}^A DE_L]^{1/2}}{2\{(1 + 2\,K_{D1}^A E_L) + [(1 + 2\,K_{D1}^A E_L)^2 + 8\,K_{D1}^A K_{D2}^A DE_L]^{1/2}\}} \tag{8}$$

The experimental values can be best fit if it is assumed that the affinity of an enzyme for the second immobilized dye is about threefold larger than for the first immobilized dye. This positive cooperativity presumably results from the high-local concentration of immobilized dye in the vicinity of the monovalent immobilized dye:enzyme complex. Again, the absolute values for these association constants are at variance with values obtained using the mobile dye, presumably resulting from the volume element considered.

We have just completed a study[33,41] of the dependence of multivalency of three dehydrogenases on the concentration of accessible immobilized Cibacron F3GA using zonal chromatography. This procedure was chosen to illuminate multivalency since the very large immobilized dye:enzyme ratio employed in zonal measurements should preferentially cause

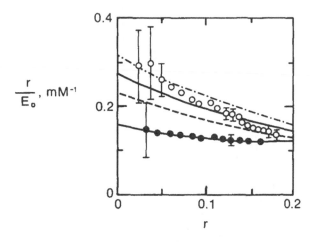

FIGURE 19. Equilibrium-partition analysis of the interaction of horse-liver alcohol dehydrogenase with Blue Sepharose. A solution of imidazole-chloride buffer, pH 7.5 and ionic strength 0.154 M maintained at 20°C, was recirculated past either 0.5 or 2.0 g of Pharmacia Blue Sepharose containing 2.5 μmol dye/mℓ. Aliquots of enzyme were added stepwise to the recirculation solution. The enzyme was allowed to equilibrate with the immobilized dye and the concentration of uncomplexed enzyme was measured at equilibrium prior to addition of the next aliquot. Results are illustrated in the form of a Scatchard plot where the binding function "r" is related to the experimental system by Equation 8. The filled circles indicate results obtained using 0.5 g of Blue Sepharose and the open circles results obtained using 2.0 g of Blue Sepharose in the recirculation measurements. The solid lines were calculated for each set of experimental results using values for $K_{D_1}^D$ and $K_{D_2}^D$ of 167 and 48 μM, respectively; the dashed line for the experimental results obtained using 2.0 g of Blue Sepharose and $K_{D_1}^D$ and $K_{D_2}^D$ values of 167 and 67 μM, respectively; and the alternatively dotted-dashed line for the experimental results obtained using 2.0 g of Blue Sepharose and $K_{D_1}^D$ and $K_{D_2}^D$ values of 167 and 37 μM, respectively. (From Hogg, P. J. and Winzor, D. J., *Arch. Biochem. Biophys.*, 240, 70, 1985. With permission.)

the enzyme to exhibit the maximal valency attainable on a column with a given geometric distribution of immobilized dye. Preliminary measurements were done to demonstrate that mass transfer was the limiting rate in these chromatographic measurements and that the proteins could exchange about eight times with each Sepharose bead before being swept on by the chromatographic flow. A series of columns were prepared having different amounts of Cibacron blue immobilized on Sepharose CL-6B. Analysis of the total immobilized-dye concentration by the method of Chambers,[29] indicated a range of values from 0.12 to 11.6 mM immobilized dye per liquid gel volume. Measurements of lactate-dehydrogenase biospecific retention indicated that 1.5 ± 0.6% of the immobilized dye was accessible to the protein on these columns. Zonal chromatographic measurements were performed using these columns and tetrameric-rabbit muscle lactate dehydrogenase, dimeric-horse liver alcohol dehydrogenase, and monomeric-scallop octopine dehydrogenase, and the mobile competitive ligands, NAD or NADH. Analysis of elution profiles using Equation 4 generated linear plots for lactate dehydrogenase using the 0.12 mM column, alcohol dehydrogenase using a 1.38 mM column, and octopine dehydrogenase using an 11.6 mM column. The dissociation constant for the immobilized dye:enzyme complex for each of these proteins was within a

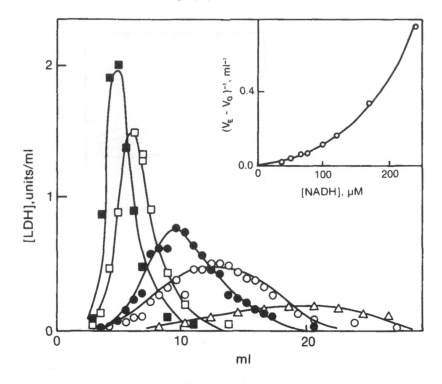

FIGURE 20. Zonal-chromatographic results obtained using lactate dehydrogenase and a Sepharose CL-6B column containing a total of 1.12 mM immobilized Cibacron blue F3GA. The column was prepared and the results obtained as described in the legend to Figure 16. The main body of the figure illustrates elution profiles obtained after zonal application of enzyme in solution containing: ■, 240 μM NADH; □, 170 μM NADH; ●, 120 μM NADH; ○, 100 μM NADH; and △, 65 μM NADH. The inset is plotted in the form of Equation 3. All measurements were obtained at ambient temperature using 100 mM phosphate buffer, pH 7.5.

factor of 2.5 of the value measured using each of these proteins and mobile dye. Zonal chromatographic measurements at higher total-immobilized dye concentrations gave curvilinear results indicative of multivalency. A typical result is shown in Figure 20. The zonal chromatographic results were analyzed for monovalency using Equation 9, which was formulated by Hethcote and DeLisi[42]

$$V_E = V_v + V_A\left[1 + \frac{DK_{D1}^A}{1 + LK_L^A}\right]$$
(9)

and is equivalent to that of Dunn and Chaiken;[37] for bivalency using Equation 10,

$$V_E = V_v + V_A\left[1 + \frac{2\,DK_{D1}^A}{1 + LK_L^A} + \frac{D^2K_{D1}^A K_{D2}^A}{(1 + LK_L^A)^2}\right]$$
(10)

which was formulated by Hethcote and DeLisi[42] and is equivalent to that of Eilat et al.[43] assuming both association constants to be identical; and for trivalency using Equation 11 and for tetravalency using Equation 12

$$V_E = V_v + V_A\left[1 + \frac{3DK_{D1}^A}{1 + LK_A^A} + \frac{3D^2K_{D1}^A K_{D2}^A}{(1 + LK_L^A)^2} + \frac{D^3K_{D1}^A K_{D2}^A K_{D3}^A}{(1 + LK_L^A)^3}\right]$$
(11)

Table 3
ANALYSIS OF ZONAL
CHROMATOGRAPHY

Enzyme	D (mM)	K_D^D (μM)	n
Lactate dehydrogenase	0.12	0.28	1.3
	0.25	0.18	2.0
	0.57	0.29	3.3
	1.12	0.30	3.0
	11.60	0.15	2.5
Alcohol dehydrogenase	1.38	5.3	1.0
	11.60	5.4	1.6

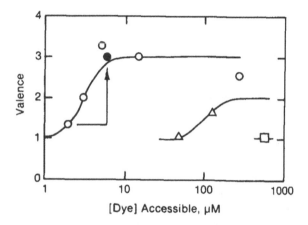

FIGURE 21. The effect of the concentration of accessible dye on the valence of immobilized dye:enzyme interactions as measured by zonal chromatography. The circles indicate measurements obtained with tetrameric-rabbit muscle lactate dehydrogenase, the triangles indicate measurements obtained with dimeric-horse liver alcohol dehydrogenase, and the square a measurement using monomeric-scallop octopine dehydrogenase. The open symbols indicate measurements obtained in 100 mM phosphate buffer, pH 7.5, and the filled symbol a measurement obtained in 10 mM phosphate buffer, pH 7.5. The bent arrow indicates the change in valence accompanying the decrease in phosphate concentration using a column having a total-immobilized dye concentration of 0.12 mM.

$$V_E = V_v + V_A\left[1 + \frac{4DK_{D1}^A}{1 + LK_L^A} + \frac{6D^2K_{D1}^A K_{D2}^A}{(1 + LK_L^A)^2} + \frac{4D^3K_{D1}^A K_{D2}^A K_{D3}^A}{(1 + LK_L^A)^3} + \frac{D^4K_{D1}^A K_{D2}^A K_{D3}^A K_{D4}^A}{(1 + LK_L^A)^4}\right] \quad (12)$$

both formulated by DeLisi and Hethcote.[44] These equations facilitate independent control of the valence and of the individual microassociation constants.

The zonal chromatographic results were first analyzed allowing the valence to vary while constraining the microassociation constant to be identical. Results of such analysis, shown in Table 3 and in Figure 21, indicate a systematic increase in valency with increasing concentration of total and accessible immobilized dye. The microdissociation constant for the immobilized dye:lactate dehydrogenase complex exhibits not only an internal consistency within the zonal chromatographic measurements of 0.24 ± 0.07 μM, but also a near identity with measurements using the mobile dye. The same comments apply to the results obtained

Table 4
ANALYSIS OF ZONAL CHROMATOGRAPHY FOR LACTATE DEHYDROGENASE

D (mM)	n	K_{D1}^D (μM)	K_{D2}^D (μM)	K_{D3}^D (μM)	K_{D4}^D (μM)	X^2 [a]
0.25	1	0.02				2.33
	2	0.20	0.14			0.05
	3	0.53	0.20	1.59		0.13
	4	4.67	0.16	−0.12	−18.0	28.0
0.57	1	0.01				21.0
	2	6.8	0.001			0.68
	3	0.24	0.23	0.19		0.28
	4	0.38	0.20	0.09	−0.95	8.2
1.12	1	0.01				98.0
	2	9.34	0.001			82.0
	3	0.22	0.35	0.25		4.2
	4	13.4	1.02	0.22	0.01	3.4
11.6	1	0.08				23.0
	2	0.20	0.02			2.5
	3	0.37	0.20	0.10		2.4
	4	0.91	0.79	0.11	0.08	2.9

[a] The term, X^2, represents the quality of the fitting procedure in terms of a chi-square value.

using alcohol dehydrogenase. Lactate dehydrogenase appears to reach a maximal observed valence of about 3 which may represent a functional maximum since a symmetrical tetramer bound simultaneously to four immobilized ligands may have no avenue of escape without substantial movement of the matrix. Alcohol dehydrogenase appears to be approaching its theoretical maximum valence of 2. However, we have not been able to construct more concentrated immobilized-dye columns to establish an observable maximal value. The same zonal chromatographic results were analyzed using Equations 9 through 12 holding the valency at integral values and allowing the microscopic association constants to vary. Results of such analyses are shown in Table 4 for lactate dehydrogenase. Within the precision of the measurements, the zonal data for lactate dehydrogenase can be fit again to a fairly uniform set of microdissociation constants and an increasing valence with increasing concentration of total immobilized dye.

Finally, our perception of the effect of salt on the concentration of accessible immobilized-Cibacron blue F3GA suggests that an immobilized dye column which is monovalent in high salt concentration could become multivalent in diminished salt concentration. Accordingly, zonal chromatographic results with lactate dehydrogenase were obtained in 10- and 100-mM phosphate buffer using a 0.12-mM immobilized-dye column. Results are shown in Figure 22. Analysis of these results indicated that the column behaved monovalently in 100-mM phosphate but trivalently in 10-mM phosphate. The change in valency is appropriate to the increase in the concentration of accessible immobilized dye as shown by the bent arrow in Figure 21.

VII. CONCLUDING REMARKS

The results summarized above indicate that reactive blue 2 immobilized at its trianzine ring retains the selectivity and affinity for protein sites characteristic for the mobile dye. This consistency not only simplifies the immobilization of the dye but also eliminates the

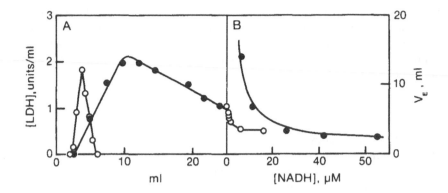

FIGURE 22.   The effect of phosphate-buffer concentration on the zonal chromatography of rabbit-muscle lactate dehydrogenase (LDH). Panel A illustrates elution profiles obtained after zonal application of enzyme to a Sepharose CL-6B column containing a total of 0.12 mM immobilized Cibacron blue F3GA equilibrated with 6 μM NADH and either, ●, 10 mM phosphate buffer or, ○, 100 mM phosphate buffer, pH 7.5. Panel B illustrates the dependence of the elution volume of the enzyme on the concentration of the competitive mobile ligand, NADH.

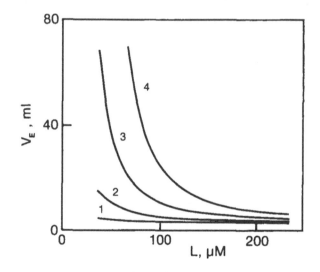

FIGURE 23.   Simulated dependence of the elution volume observed by zonal chromatography on the valence of the protein. Simulations were done using Equations 9 to 12 and K_{D1}^D to K_{D4}^D of 0.3 μM, K_L^P of 3.4 μM and a concentration of accessible immobilized Cibacron blue F3GA of 15.1 μM. The numbers 1, 2, 3, and 4 indicate the valence.

potential perturbations in selectivity and affinity introduced by spacer groups. The preliminary results of Lowe and coworkers[28] indicate that the structure of the dye can be tailored by relatively simple chemistry to more selectively retain a desired protein. Frontal-chromatographic, zonal-chromatographic, and equilibrium-partition measurements have all shown that multivalent interactions between immobilized dyes and oligomeric proteins can occur if the concentration and density of accessible immobilized dye is appropriate. If this is the case, multivalency can be a powerful selective parameter among proteins having the same affinity for immobilized dye but different multivalent potential as shown schematically in Figure 23. Salt concentration appears to be a significant variable in these considerations, effecting both the capacity and multivalency of a column by virtue of controlling the concentration and density of the immobilized dye. It is anticipated that future quantitative measurements of the interactions between immobilized dyes and proteins will be attentive to the source of the dye and the reaction conditions for immobilization so that matrices will contain an immobilized dye of known structure randomly distributed.

VII. GLOSSARY OF TERMS

D	Concentration of accessible dye
D_o	Concentration of total dye
E_L	Concentration of mobile enzyme at equilibrium
E_M	Concentration of enzyme in the matrix at equilibrium
E_o	Initial concentration of enzyme
f	Number of equivalent sites per enzyme oligomer
K	Partition coefficient
K_D^A	Association constant for dye:enzyme complex; subscripts 1, 2, 3, or 4 refer to individual microscopic association constants in multivalent complexes
K_D^D	Dissociation constant for dye:enzyme complex; subscripts 1, 2, 3, or 4 refer to individual microscopic dissociation constants in multivalent complexes
K_L^A	Association constant for mobile competitive ligand:enzyme complex
K_L^D	Dissociation constant for mobile competitive ligand:enzyme complex
L	Concentration of mobile competitive ligand at equilibrium
L_o	Initial concentration of mobile competitive ligand
n	Valence
V	Volume in which enzyme is introduced
V_A	Volume of matrix accessible to enzyme
V_E	Volume in which enzyme is eluted; maximum ordinate of elution peak
V_i	Difference in the elution volume for materials having a partition coefficient of 1 and 0
V_L	Same as V_E
V_o	Elution volume for enzyme from equivalent unconjugated matrix
V_V	Elution volume for material having a partition coefficient of 0

REFERENCES

1. **Dean, P. and Quadri, F.,** Affinity chromatography on immobilized dyes, in *Solid Phase Biochemistry,* Scouten, W. H., Ed., Wiley-Interscience, New York, 1983, 79.
2. **Kopperschläger, G., Böhme, H-J., and Hofmann, E.,** Cibacron blue F3G-A and related dyes as ligands in affinity chromatography, *Adv. Biochem. Eng.,* 25, 101, 1982.
3. **Lowe, C. R., Small, D. A. P., and Atkinson, A.,** Some preparative and analytical applications of triazine dyes, *Int. J. Biochem.,* 13, 33, 1981.
4. **Subramanian, S.,** Dye-ligand affinity chromatography: the interaction of cibacron F3GA with proteins and enzymes, *CRC Crit. Rev. Biochem.,* 16, 169, 1984.
5. **Dean, P. D. G. and Watson, D. H.,** Protein purification using immobilised triazine dyes, *J. Chromatogr.,* 165, 301, 1979.
6. **Baksi, K., Rogerson, D. L., and Rushizky, G. W.,** Rapid, single-step purification of restriction endonucleases on cibacron blue F3GA-agarose, *Biochemistry,* 17, 4136, 1978.
7. **George, J. and Chirikjian, J. G.,** Biospecific fractionation matrices for sequence specific endonucleases, *Nucl. Acids Res.,* 5, 2223, 1978.
8. **Sobhanaditya, J. and Rao, N. A.,** Plant flavokinase: interaction of cibacron blue F3GA with the enzyme from *Vigna radiata* (mung bean) seeds, *Biochem. Int.,* 5, 567, 1982.

9. **Lascu, I., Duc, M., and Cristea, A.,** Rapid large-scale purification of pig heart nucleoside diphosphate kinase by affinity chromatography on cibracon blue 3G-A sepharose, *Anal. Biochem.*, 113, 207, 1981.

10. **Thompson, S. T., Cass, K. H., and Stellwagen, E.,** Blue dextran-sepharose: an affinity column for the dinucleotide fold in proteins, *Proc. Natl. Acad. Sci. U.S.A.*, 72, 669, 1975.

11. **Ramadoss, C. S., Steczko, J., Uhlig, J. W., and Axelrod, B.,** Effect of albumin on binding and recovery of enzymes in affinity chromatography on cibracon blue, *Anal. Biochem.*, 130, 481, 1983.

12. **Lascu, I., Porumb, H., Porumb, T., Abrudan, I., Tarmure, C., Petrescu, I., Presecan, E., Proinov, I., and Telia, M.,** Ion-exchange properties of cibacron blue 3G-A sepharose (blue sepharose) and the interaction of proteins with cibacron blue 3G-A, *J. Chromatogr.*, 283, 199, 1984.

13. **Robinson, J. B., Jr., Strottmann, J. M., Wick, D. G., and Stellwagen, E.,** Affinity chromatography in nonionic detergent solutions, *Proc. Natl. Acad. Sci. U.S.A.*, 77, 5847, 1980.

14. **Ashton, A. R., and Polya, G. M.,** The specific interaction of cibacron and related dyes with cyclic nucleotide phosphodiesterase and lactate dehydrogenase, *Biochem. J.*, 175, 501, 1978.

15. **Bruton, C. J., and Atkinson, T.,** The binding of aminoacyl-tRNA synthetases to triazine dye conjugates, *Nucl. Acids Res.*, 7, 1579, 1979.

16. **Lowe, C. R., Hans, M., Spibey, N., and Drabble, W. T.,** The purification of inosine 5'-monophosphate dehydrogenase from *Escherichia coli* by affinity chromatography on immobilized procion dyes, *Anal. Biochem.*, 104, 23, 1981.

17. **Kopperschläger, G., Freyer, R., Diezel, W., and Hofmann, E.,** Some kinetic and molecular properties of yeast phosphofructokinase, *FEBS Lett.*, 1, 137, 1968.

18. **Haeckel, R., Hess, B., Lauterborn, W., and Wuster, K-H.,** Purification and allosteric properties of yeast pyruvate kinase, *Hoppe-Seyler's Z. Physiol. Chem.*, 349, 699, 1968.

19. **Weber, B. H., Willeford, K., Moe, J. G., and Piszkiewicz, D.,** Hazards in the use of cibacron blue F3GA in studies of proteins, *Biochem. Biophys. Res. Commun.*, 86, 252, 1979.

20. **Issaly, I., Poiret, M., Tauc, P., Thiry, L., and Herve, G.,** Interactions of cibacron blue F3GA and nucleotides with *Escherichia coli* aspartate carbamoyltransferase and its subunits, *Biochemistry*, 21, 1612, 1982.

21. **Seelig, G.,** personal communication cited in Reference 22.

22. **Hanggi, D. and Carr, P.,** Analytical evaluation of the purity of commercial preparations of cibacron blue F3GA and related dyes, *Anal. Biochem.*, 149, 91, 1985.

23. **Böhme, H. J., Kopperschläger, G., Schultz, J., and Hofmann, E.,** Affinity chromatography of phosphofructokinase using cibacron blue F3G-A, *J. Chromatogr.*, 69, 209, 1972.

24. **Heyns, W. and de Moor, P.,** Conversion of 5α-Dihydrotestosterone into 5α-androstane-3β,17β-diol and purification of the enzyme by affinity chromatography, *Biochim. Biophys. Acta*, 358, 1, 1974.

25. **Baird, J. K., Sherwood, R. F., Carr, R. J. G., and Atkinson, A.,** Enzyme purification by substrate elution chromatography from procion dye-polysaccharide matrices, *FEBS Lett.*, 70, 61, 1976.

26. **Thompson, S. T. and Stellwagen, E.,** Binding of cibacron blue F3GA to proteins containing the dinucleotide fold, *Proc. Natl. Acad. Sci. U.S.A.*, 73, 361, 1976.

27. **Biellmann, J-F., Samama, J-P., Brandén, C. I., and Eklund, H.,** X-ray studies of the binding of cibacron blue F3GA to liver alcohol dehydrogenase, *Eur. J. Biochem.*, 102, 107, 1979.

28. **Lowe, C. R., Burton, S. J., Pearson, J. C., Clonis, Y. D., and Stead, C. V.,** Design and application of bio-mimetic dyes in biotechnology, *J. Chromatogr.*, 376, 121, 1986.

29. **Chambers, G. K.,** Determination of cibacron blue F3GA substitution in blue sephadex and blue dextran-sepharose, *Anal. Biochem.*, 83, 551, 1977.

30. **Thresher, W. C. and Swaisgood, H. E.,** On the preparation and binding characteristics of cibacron blue-glycerolpropyl-glass, *Biochim. Biophys. Acta*, 749, 214, 1983.

31. **Hogg, P. J. and Winzor, D. J.,** Quantitative affinity chromatography: further developments in the analysis of experimental results from column chromatography and partition equilibrium studies, *Arch. Biochem. Biophys.*, 234, 55, 1984.

32. **Liu, Y. C., Ledger, R., and Stellwagen, E.,** Quantitative analysis of protein: immobilized dye interaction, *J. Biol. Chem.*, 259, 3796, 1984.

33. **Liu, Y-C.,** Quantitative Analysis of Protein:Immobilized Dye Interaction, Ph.D. thesis, University of Iowa, 1985.

34. **Robinson, J. R., Jr., Strottmann, J. M., and Stellwagen, E.,** Prediction of neutral salt elution profiles for affinity chromatography, *Proc. Natl. Acad. Sci. U.S.A.*, 78, 2287, 1981.

35. **Taylor, J. B. and Swaisgood, H. E.,** A unified partition coefficient theory for chromatography, immobilized enzyme kinetics and affinity chromatography, *Biotechnol. Bioeng.*, 23, 1349, 1981.

36. **Nichol, L. W., Ogston, A. G., Winzor, D. J., and Sawyer, W. H.,** Evaluation of equilibrium constants by affinity chromatography, *Biochem. J.*, 143, 435, 1974.

37. **Dunn, B. M. and Chaiken, I. M.,** Quantitative affinity chromatography. Determination of binding constants by elution with competitive inhibitors, *Proc. Natl. Acad. Sci. U.S.A.*, 71, 2382, 1974.

38. **Graves, D. J. and Wu, Y. T.**, On predicting the results of affinity procedures, *Methods Enzymol.*, 34, 140, 1974.
39. **Yon, R. J. and Kyprianou, P.**, Biospecific desorption from low-specificity adsorbents with emphasis on 10-carboxydecylamino-sepharose, in *Affinity Chromatography and Biological Recognition*, Chaiken, I. M., Wilchek, M., and Parikh, I., Eds., Academic Press, Orlando, 1983, 143.
40. **Hogg, P. J. and Winzor, D. J.**, Effects of solute multivalency in quantitative affinity chromatography: evidence for cooperative binding of horse liver alcohol dehydrogenase to blue sepharose, *Arch. Biochem. Biophys.*, 240, 70, 1985.
41. **Liu, Y. C. and Stellwagen, E.**, Zonal chromatographic analysis of the interaction of alcohol dehydrogenase with blue sepharose, *J. Chromatogr.*, 376, 149, 1986.
42. **Hethcote, H. W. and DeLisi, C.**, Determination of equilibrium and rate constants by affinity chromatography, *J. Chromatogr.*, 248, 183, 1982.
43. **Eilat, D. and Chaiken, I. M.**, Expression of multivalency in the affinity chromatography of antibodies, *Biochemistry*, 18, 790, 1979.
44. **DeLisi, C. and Hethcote, H. W.**, Chapter 1, this text.

Appendix

NOTATIONS USED FOR PARAMETERS AND CHARACTERISTICS OF ANALYTICAL AFFINITY CHROMATOGRAPHY

	Symbols used			
Definition	**DeLisi -Hethcote**	**Swaisgood -Chaiken**	**Walters**	**Stellwagen -Liu**
Molecular Interactions				
Matrix-immobilized interactant (affinity matrix)		M	L	
Concentration of un-liganded-immobilized interactant		[M]		
Moles of immobilized interactant in column			m_L	
Total concentration of accessible-immobilized interactant	N	$[M]_T$	m_L/V_p	D
Concentration of free sites of immobilized interactant	N*			
Mobile interactant (protein, enzyme, ligand, etc.)		P,L	E	
Total concentration mobile interactant		$[P_T],[L]_T$		E_1, or E_m
Total-initial concentration of mobile interactant		$[P]_T^o,[L]_T^o$		E_o
Valency of multivalent-mobile interactant		f		n
Soluble-interacting competitor (or other effector)		L	I	
Concentration of soluble-interacting competitor (or other effector)	L	[L]	[I]	L
Moles of soluble-interacting molecule bound per mole of immobilized interactant		$\bar{\nu}$		
Equilibrium-Interaction Constants				
Equilibrium-dissociation constant of binary complex of molecules I and J		$K_{1,J}$		

Appendix (continued)

NOTATIONS USED FOR PARAMETERS AND CHARACTERISTICS OF ANALYTICAL AFFINITY CHROMATOGRAPHY

Definition	Symbols used			
	DeLisi -Hethcote	Swaisgood -Chaiken	Walters	Stellwagen -Liu
Equilibrium-disso-ciation constant of binary complex of immobilized and mobile interactants	$1/K_3$	$K_{M/P}, K_{M/L}$	$1/K_3$	$K_D^D = (1/K_D^A)$
Equilibrium-disso-ciation constant of binary complex of mobile interactant and soluble competitor	$1/K_2$	$K_{P/L}$	$1/K_2$	$K_L^D = (1/K_L^A)$
Equilibrium-associa-tion constants of second, third, and fourth binding of mobile interactant to immobilized interactant	K_4, K_5, K_6			
Equilibrium-disso-ciation constant of a ternary complex IJK, into IJ and K, in which dissocia-tion of molecule K is from molecule J in the complex		$K_{IJ^*.K}$		
Intrinsic equilib-rium-association constant of the "i"th class of sites for a molecule ex-hibiting multiple-binding equilibria		k_i		
Subscripts referring to microassociation and microdissocia-tion constants of individual-mobile interactant: matrix complexes for oli-gomeric-mobile interactant				$1,2,3,\ldots$
Elution Characteristics				
Elution volume of mobile interactant on affinity matrix	V_e	V	V_r	V_e or V_1

Appendix (continued)

NOTATIONS USED FOR PARAMETERS AND CHARACTERISTICS OF ANALYTICAL AFFINITY CHROMATOGRAPHY

Definition	Symbols used			
	DeLisi -Hethcote	Swaisgood -Chaiken	Walters	Stellwagen -Liu
Elution volume of unretarded molecule on affinity matrix (void volume)		V_o	V_m	V_o
Volume of the mobile phase	V_o	V_m	V_e	V_v
Volume of the stationary phase (accessible pore or penetrable volume in the case of porous matrix)	V_p	V_S	V_p	V_i
Peak position in time units for mobile interactant	M_e	$E(t')$	t_r	
Peak position in time units for unretarded-mobile molecule		$E(t_o)$	t_m	
Equilibrium-partition coefficient of mobile molecule interacting with affinity matrix		σ_i		
Capacity factor			k'	
Equilibrium-partition coefficient for a molecule unretarded on affinity matrix		$\sigma_{o,i}$		
Partition coefficient of mobile interactant with matrix in continuous-elution chromatography		K_{AV}		
Standard deviation of peak of unretarded-mobile molecule		σ_o		
Standard deviation of peak of mobile interactant		σ'	σ	
Variance of eluted peak of mobile interactant, in time/volume units	S_e/W_e	$(\sigma')^2/W_e$	$\sigma^2/-$	

Appendix (continued)

NOTATIONS USED FOR PARAMETERS AND CHARACTERISTICS OF ANALYTICAL AFFINITY CHROMATOGRAPHY

	Symbols used			
Definition	**DeLisi -Hethcote**	**Swaisgood -Chaiken**	**Walters**	**Stellwagen -Liu**
Third central moment of time-elution profile	N_e			
Third central moment of volume-elution profile	U_e			
Moment coefficient of skewness	G_1			
First moment of advancing boundary of mobile interactant in continuous-elution chromatography		$\bar{V}$		
Limiting value for $\bar{V}$ as concentration of mobile interactant approaches zero		$\bar{V}_{lim}$		
Denotation of stationary phase		α		
Denotation of mobile phase		β		

Rate Constants for Interactions, Other Processes

Kinetic association/dissociation rate constant for mobile interactant to affinity matrix	k_3/k_{-3}	k_a/k_d	k_3/k_{-3}	
Kinetic-association/dissociation rate constant, for mobile interactant to soluble competitor	k_2/k_{-2}		k_2/k_{-2}	
Mass-transfer adsorption/desorption rate constant, for solute into and out of stagnant-mobile phase	k_1/k_{-1}	k_1/k_{-1}	k_1/k_{-1}	
First order adsorption-rate constant			k_3^*	
Association-rate constant for second, third, and fourth binding of mobile interactant to affinity matrix	k_4, k_5, k_6			

Appendix (continued)

NOTATIONS USED FOR PARAMETERS AND CHARACTERISTICS OF ANALYTICAL AFFINITY CHROMATOGRAPHY

	Symbols used			
Definition	**DeLisi -Hethcote**	**Swaisgood -Chaiken**	**Walters**	**Stellwagen -Liu**
Dissociation rate for second, third, and fourth binding of mobile interactant to affinity matrix	k_{-4}, k_{-5}, k_{-6}			
Diffusion constant of mobile interactant	D		D_m	
Chromatographic System Characteristics				
Volumetric-flow rate	F	F	F	
Linear velocity of an excluded solute (flow velocity)	u		u_e	
Linear velocity of unretarded-mobile molecule			u	
Column height or length	h	h	L_{col}	
Empty-column volume			V_{col}	
Surface area of affinity matrix			A	
Matrix-particle diameter	d_p		d_p	
Tortuosity			γ	
Sample thickness	T	T		V
Plate height	H		H	
Number of plates			N	

INDEX

A

Accessible immobilized dye, 164—170, 175, 177, 179

Affinity chromatography, see also Analytical affinity chromatography; Quantitative affinity chromatography, 3—4, 13, 29, 58—61
  central goal in design of, 66
  rate constant measurement by, 117—156
    isocratic method, 121—136
    peak-decay method, 145—155
    split-peak method, 137—145

Analytical affinity chromatography, see also specific topics, 65—115
  continuous elution, 68, 71, 80, 84, 86—92
  glossary of terms, 109—113
  mobile component-immobilized component interaction kinetics, 92—94
  molecular interactions, summary of equations for analysis of, 104—109
  multiple interactions, 94—112
    analysis, 104—108
    characterization, 94—100
  neurophysin/neurohypophysial hormone system interactions, 94—104
  zonal elution, 67—86
    bivalent systems, 71
    monovalent systems, 67—70

Antibody-antigen complexes, 138

Apolar molecules, 12—13

Association, 35

Association constant, 91, 176

Association rate constant, 50, 58, 92, 137—145

Average binding numbers, 54

Axial diffusion, see also Diffusion, 15, 18, 20—21, 25, 119

B

Band broadening, 33, 118—121, 123—124, 131, 137, 148

Band width, 31

Bead-bound ligand, 7—9, 35, 42, 46, 49, 52

Bead-bound macromolecules, 49—51, 53, 55, 57—58, 61

Bead-bound molecule, 8

Beads, 5—6, 8, 10
  cross-linked polymers, 13
  diameter, 33—34, 57
  heterogeneity, 22
  impenetrable, 46—49, 55, 57, 60—61
  nonporous, 76—77, 93—95, 121
  particle size, 33
  surface, 10
  transition layer surrounding, 7
  uniformity, 25

Bead-sized heterogeneity, 14

Bessel functions, 13, 16

Biomimetic dye, see also Immobilized dyes, 162

Bivalent binding, 42—45, 48—49, 52, 55, 59—61, 71—72

Bivalent interactions, 104—105

Boundary layer thickness, 7

Bovine serum albumin, 90—91

Break-through analysis, 132—133, 142

Brownian diffusion, 17

Brownian motion, 4, 6—8, 10

Bulk phase, see Mobile phase

Bulk solution, 9—10

C

Capacity factor, 121—123, 149

Cell-bound receptors, 3

Central potential, 8

Chemical reaction dissociation constants, 56

Chemical reaction kinetics, see Chemical reactions

Chemical reaction rate constant, estimation, 55—58

Chemical reactions, 3—4, 7, 13, 33, 38, 50—52, 57, 93

Chromatographic system characteristics, 189

Cibacron blue, see also Immobilized dyes, 158, 160

Cibacron blue complex, 162

Column-scanning chromatography, 31—33

Competitive elution analysis, 104—105

Computer simulations
  isocratic method, 125—127
  peak-decay method, 147—151
  split-peak method, 138—139

Concanavalin A-sugar kinetics, 129—130, 141, 151—152

Constant velocity assumption, 7

Continuous elution analytical affinity chromatography, 20, 31, 68, 71, 80, 84, 86—92
  bovine serum albumin, 90—91
  glossary of terms, 110
  immobilized molecule, 107
  monovalent interactant, 105
  multivalent components, 91—92
  trypsin-ligand interactions, 86—90

Convection, 17—18, 33

Cooperativity, 175, 176

Coulomb potential, 12

Coulomb's law, 10—11

D

Density functions, 15

Desorption, 118—119, 148—149, 151

Desorption rate constant, 14, 31, 34, 38, 49—50, 57—58, 93

Dialysis, 4

Dielectric constants, 12

Printed and bound by CPI Group (UK) Ltd, Croydon, CR0 4YY

22/10/2024

01777632-0004